WNA

GAS CHROMATOGRAPHY–MASS SPECTROMETRY IN NEUROBIOLOGY

Advances in Biochemical Psychopharmacology

Volume 7

Advances in Biochemical Psychopharmacology

Series Editors

Erminio Costa, M.D.

Chief, Laboratory of Preclinical Pharmacology
National Institute of Mental Health
Washington, D.C., U.S.A.

Paul Greengard, Ph.D.

Professor of Pharmacology
Yale University School of Medicine
New Haven, Connecticut, U.S.A.

GAS CHROMATOGRAPHY–MASS SPECTROMETRY IN NEUROBIOLOGY

Advances in Biochemical

Psychopharmacology

Volume 7

EDITORS

E. Costa, M.D.

Chief, Laboratory of Preclinical
Pharmacology
National Institute of Mental Health
Washington, D.C., U.S.A.

B. Holmstedt, M.D.

Department of Toxicology
Swedish Medical Research Council
Karolinska Institutet
Stockholm, Sweden

Raven Press ▪ New York

International Standard Book Number 0–911216–48–0
Library of Congress Catalog Card Number 73–84113

Preface

It is a privilege to be asked to write a preface to so significant a book. It is also stimulating: for when one has long worked with conventional methods, with all their limitations, it catches the imagination to see what had seemed major barriers to advance suddenly melting. The conjoined use of gas chromatography, mass spectrometry and appropriate computer aid (for control of the spectrometer as well as for data processing) seems to offer a quantal change in technique comparable to that of the electron microscope. Where one used to struggle to detect micrograms of some biologically interesting substance, now picograms become possible, and areas beyond this are opening up. Nor is this sensitivity narrow; already a substantial mass range is covered. The specificity achievable, for the pharmacologist who has had to struggle with parallel assay, is remarkable and sometimes almost redundant.

One is conscious, too, that public attention is turning increasingly to the problems presented by small quantities of chemical substances in the internal and external environments; one can only be grateful, therefore, for the development work that has produced a technology capable of meeting the new challenge—almost, so it seems, at the crucial moment.

The Editors have given the reader much to be grateful for. The papers cover a wide range, involving a variety of instruments, showing the power of the method in measuring minute amounts of drugs and their metabolites; and of neurotransmitters and 'false transmitters' and their metabolites; in measurements on very small, defined amounts of tissue; and in determining chemical structure. While it is invidious to pick out any special study, many readers will be disconcerted by the findings, made possible by the techniques, that a newborn child may contain in its body a narcotic analgesic, a barbiturate and caffeine. No such compilation can hope to stay in date: yet the book is worth reading thoroughly, not so much for the findings of the moment, but for the experimental and conceptual doors it opens. The Editors earn a double debt of gratitude by collating the available references at the end—a list which itself stimulates one's experimental imagination not only to do old types of experiments better, but also to make new investigations in new ways.

W.D.M. Paton
University Department of Pharmacology
Oxford

Contents

Advances in Biochemical Psychopharmacology, Vol. 7
Raven Press, New York © 1973

Mass Fragmentography: Principles, Advantages, and Future Possibilities

Bo Holmstedt and Lena Palmér

*Department of Toxicology, Swedish Medical Research Council, Karolinska Institute,
S–104 01 Stockholm 60, Sweden*

I. GAS CHROMATOGRAPHY–MASS SPECTROMETRY (GC–MS)

The combination of a gas chromatograph (GC) with a mass spectrometer (MS) permits direct identification of compounds. The advantage of this technique is the combination of great separation power (GC) with great identification power (MS). The usefulness of this technique in research on drug metabolism is now generally recognized. The identity of a compound is established by comparing its mass spectrum to that of a reference substance. Elucidating the structure of a total unknown may be difficult, but with modern drugs, the chemical structure is always known and synthetic analogues are usually available. Knowledge of mass spectra and fragmentation patterns forms the basis for the elucidation of the structure of both drug metabolites and endogenous compounds. Such a knowledge is also a *sine qua non* for the use of the techniques called *mass fragmentography*.

The principles and application of the combination of gas chromatography and mass spectrometry in biomedical sciences have been reviewed in detail by Hammar, Holmstedt, Lindgren, and Tham (1969). More recent reviews are those of Hammar (1971), Holmstedt and Linnarson (1972), Waller (1972), and Jenden and Cho (1972). This chapter will deal only with the use of the mass spectrometer as an ion specific detector in biomedical analysis.

II. DEFINITION AND GENERAL PRINCIPLES

In the magnetic mass spectrometer the ions will follow a circular pathway and only ions of a certain m/e value will be focused on the detector for a given accelerating voltage (V) and a given magnetic field (H). The following formula gives the m/e value for an ion in focus:

$$m/e = \frac{H^2 \cdot R^2}{2V}$$

where R is the radius of the ion beam. By rapid switching between selectable values of V, keeping the magnetic field constant, ions of the desired m/e values may consecutively be brought into focus (leaving other ions undetected) (Fig. 1).

Using ion specific detection, the intensity (abundance) of the ions produced when the effluent from the gas chromatograph is ionized in the ion source, may be recorded as continuous curves. Depending upon the equipment used, different numbers of fragments typical of the compound to be analyzed may be detected simultaneously. A mass fragmentogram is similar in appearance to a gas chromatogram. Like the latter it gives the following information: the retention time which is characteristic of the compound and the peak area (of any of the peaks) which is proportional to the amount. In addition, the relative intensities of the peaks should be the same as the relative intensities of the ionic fragments in the mass spectrum of the compound. This is another criterion of identity.

Mass fragmentography is the exclusive use of the spectrometer for the mass specific recording of preselected ions. A mass fragmentogram is obtained by on-line analogue processing of data, and requires that the mass

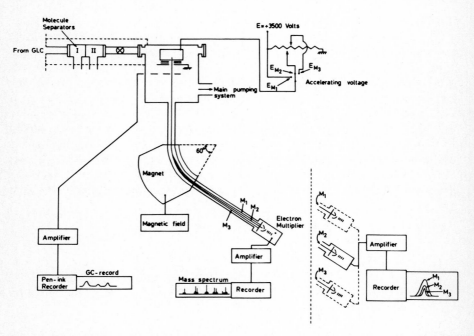

FIG. 1. Schematic diagram of mass spectrometer equipped with multiple-ion detector.

spectrometer be focused beforehand on the masses at which the ion abundance are to be measured (Hammar, Holmstedt, and Ryhage, 1968b). By contrast to mass fragmentography, *mass chromatography* is the analysis of data from repetitively scanned mass spectra that are stored for the most part on magnetic tape or disc to enable recovery of a graph of a specific ion abundance against time. No previous knowledge of the masses of interest is required. Any masses can be recalled (Hites and Biemann, 1970). The sensitivity is considerably lower than that of mass fragmentography.

III. DEVELOPMENT OF MASS FRAGMENTOGRAPHY

The tracing of a single ion appearing during a gas chromatographic run was carried out even before the arrival of molecule separators. This approach was used in the past by Henneberg (1959 and 1961), Gohlke (1962), Henneberg and Schomburg (1962). Sometimes a gas chromatogram was recorded simultaneously, and quantitation was attempted (Henneberg, 1961). Single-channel mass fragmentography later became useful in qualitative work on flavor components (Kolor, 1972) and has been used quantitatively for the determination of transmitter substances in biological material (Cattabeni, Koslow, and Costa, 1972).

In 1960 Lindeman and Annis made use of the direct combination of the mass spectrometer and the gas chromatograph for the analysis of volatile organic mixtures. Their technique consisted of repetitive scanning and subsequent manual measurement and evaluation of the relative abundance of certain mass numbers. In this way they reconstructed what is now called mass chromatograms. The wealth of data obtained by Lindeman and Annis in a single analysis took about a week of calculating time to process. Strangely enough it was the paper by Lindeman and Annis which gave the impulse to Sweeley and his co-workers to construct the accelerating voltage alternator (AVA) (Sweeley, Elliot, Fries, and Ryhage, 1966). They published a technique in which two accelerating voltages were alternated so as to resolve those components that had been eluted simultaneously from the gas chromatograph. Two mass numbers were recorded. The method was used to determine the isotope abundance in a mixture of penta-O-trimethylsilyl derivatives of β-glucose and β-glucose-D_7 and in mixtures of the trimethylsilyl derivative of epiandrosterone and dehydroepiandrosterone. Amounts as small as 20 ng of the compounds were detected.

The work by Sweeley et al. (1966) suggested the idea of simultaneous mass fragmentography of drugs and metabolites. This technique was reported by Hammar et al. (1968b) in a paper showing its application to the identification of chlorpromazine and some of its metabolites in human blood.

By continuous recording of m/e values corresponding to characteristic fragments of the expected metabolites it was possible to eliminate or at least diminish the influence of interfering compounds. Characterization of the metabolites was based upon the presence of selected mass numbers, their relative intensities, and retention times. Thus, the mass spectrometer was used as a gas chromatographic detector with improved selectivity compared to single-ion detection. This technique was so sensitive that picogram quantities (injected amounts) or 10^{-14} to 10^{-15} moles could be determined.

IV. EQUIPMENT AVAILABLE

A. Accelerating Voltage Alternator (AVA)

The LKB AVA-accessory has three channels. The lowest mass (m/e) was focused on the first channel at basic accelerating voltage by adjustment of the magnetic field. By reducing the accelerating voltage by means of resistors, higher masses, which do not exceed the lowest mass by more than 10%, can be focused on the two remaining channels. The alternations between the different channels are recorded with the same galvanometer and give rise to a curve difficult to interpret (Hammar et al., 1968b).

B. Multiple-Ion Detector (MID)

The MID described by Hammar and Hessling (1971) had a mass range of 20% of the lowest mass within which three different masses could be focused. The required accelerating voltages (preset by potentiometers and measured by a voltmeter) to bring the selected masses into focus were added to the basic accelerating voltage. The curves for individual masses were made continuous by the utilization of a sample-hold circuit for each mass channel and by using one galvanometer for each channel. Individual channel controls for the amplification of the ion signals and suppression of the background signals were introduced, allowing compensation for large differences in the intensities of the selected masses and elimination of the difficulties caused by the column bleeding. These facilities permitted the use of higher sensitivities.

A second MID described by Hammar (1972) allows a digital setting of the channel voltages by thumb-wheel-switchers and a D/A-converter. This MID has four channels. The mass range is increased approximately to 30% of the lowest mass. There is a flexibility with regard to the masses which can be focused within this range.

C. Multiplexer

Devices have been described for use with the quadrupole instruments which allow up to eight masses to be recorded over the entire mass range, with separate demodulated outputs (Wiesendanger and Tao, 1970; Bonelli, 1972; Cho, Lindeke, Hodson, and Jenden, 1972; Jenden and Silverman, 1973).

D. Computer Controlled-Accelerating Voltage and Signal Treatment

A flexible GC-MS laboratory computer system has recently been developed for on-line data collection and processing (Elkin, Pierrou, Ahlborg, Holmstedt, and Lindgren, 1973). A program to sample simultaneously four masses and record the resulting data is part of this system. The data is displayed on the scope of a PDP 12 computer continually and all questions and answers are via the scope. This allows quick evaluation of data that may be saved on tape, printed, or plotted. After collection of this data the peak areas and the ratios of peak areas are displayed. A plot is shown in Fig. 2. Control of the accelerating voltage of the mass spectrometer is carried out with a D/A converter. All signal processing is confined to the computer. Digital filters are now used instead of the analogue filters in previous systems. The hardware consists of a voltage control unit and one base-line and gain control unit used sequentially for all channels. In the previously used mid-systems gain and base-line controls were manually adjusted with separate units, one for each channel. The mass range is still approximately 30% (calculated on the lowest mass).

The advantage of the system is that it is not limited to hardware implemented channels as in the previously used ones. If additional channels are required, extra channel time is included in the software.

V. QUALITATIVE MASS FRAGMENTOGRAPHY

Mass fragmentography was first used to advantage in studies of the basic metabolites of chlorpromazine in plasma (Hammar et al., 1968*b*). Regular gas chromatography-mass spectrometry could successfully be used only with an acidic metabolite occurring in rather high concentrations in the blood of patients. The method failed when basic metabolites were to be analyzed. The mass spectra obtained could not be interpreted because of the low concentration and the interference from other components in the

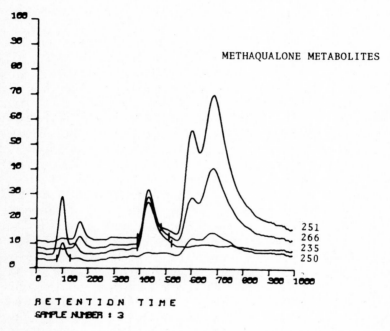

SAMPLE NO. :   3

CHANNEL	MASS	AREA	RET.TIME	RATIO
1	251	3519	215	.584648
2	266	6019	215	1
3	235	5913	215	.982389
4	250	760	49	.126266

FIG. 2. Ion-specific recordings of an extract of urine from a patient receiving methaqualone.

extract. However, using the characteristic fragments of the ring skeleton, 2-chloro-phenothiazinyl, and the fragment with one methylene group attached to the ring nitrogen, it was possible to screen for a whole group of nonring-substituted 2-chloro-phenothiazinyl compounds.

Two metabolites of chlorpromazine, monodesmethyl- and didesmethylchlorpromazine, were easily recognized by their retention times and by the relative intensities of the typical mass numbers. By refocusing, the molecular ion and its isotope were also used for detection. A complete agreement existed between the reference compounds and the metabolites with regard to the focused fragments. When these requirements are fulfilled it can be firmly stated that there is identity between the references and the compounds. None of the mass numbers used gave rise to peaks when plasma from people not given chlorpromazine was treated in the same way. Tricyclic antidepressants [nortriptyline (NT) and some of its metabolites] behave in mass spectrometry similarly to the chlorpromazine metabolites in that nonderivatives give very poor spectra in the higher mass region. After conversion to amides, the heavier fragments, having the tricyclic ring system as a constituent, dominate. These are better suited for mass fragmentography (for references to this work see Hammar, 1971).

It was possible by focusing upon mass numbers 217, 219, and 232 to detect nonring-substituted metabolites as well as the parent drug. When these mass numbers were used to analyze a basic and trifluoroacetylated extract of cerebrospinal fluid from a NT-intoxicated person, two large peaks showed up in the mass fragmentogram. The second one with a retention time similar to NT-TFA contained just one mass number, 217 (Fig. 3). The evidence strongly suggested that this substance was a metabolite of NT: 1) it had a similar retention time; 2) it was present in significant amounts; 3) it had at least one mass number in common with NT-TFA; and 4) it was isolated by the same procedures.

An inspection of the mass spectrum of NT-TFA indicated that the fragment m/e 217 might result from the loss of two hydrogen atoms in the etylene bridge between the two benzene rings, forming a fully conjugated and stable ring system. This theory would fit well with the absence of mass numbers 219 and 232. If these hypotheses were correct, the mass numbers 202, 215, 217, 230, and 357 should be present whereas those typical of NT-TFA, 204, 219, 232, and 359 should be lacking. This was confirmed by MF with refocusing. Final confirmation came with the synthesis of dehydro-NT (DH_2NT), which gave a mass spectrum identical to that of the second peak. There were reasons to believe that DH_2NT could be formed *in vitro* by the loss of H_2O from 10-hydroxynortriptyline (10-OH-NT). When synthetic 10-OH-NT became available, the final proof that 10-OH-NT is a metabolite was obtained. There is no disadvantage in determining 10-OH-NT by means of its derivatized form DH_2NT-TFA, if it can be clearly demonstrated that DH_2NT is not a constituent of the sample (Hammar, Alexanderson, Holmstedt, and Sjöqvist, 1971). 10-oxo-NT was recently found by mass fragmentography when 10-OH-NT was incubated with rat liver microsomes (von Bahr, 1972).

Drugs labeled with stable isotopes can be used for the identification of metabolites in body fluids (Knapp and Gaffney, 1972; Knapp, Gaffney, and McMahon, 1972*a*; Knapp, Gaffney, McMahon, and Kiplinger, 1972*b*). Many other examples of qualitative mass fragmentography are available (see selected list of references elsewhere in volume).

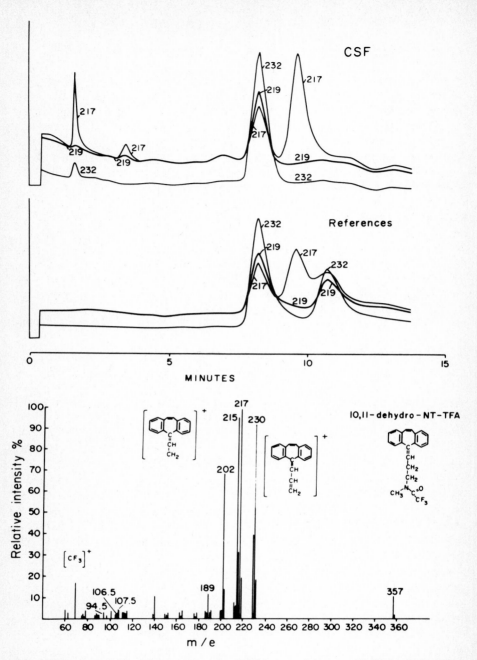

FIG. 3. Upper panel: Mass fragmentogram of TFAA-treated extract of CSF drawn from an intoxicated person. References of NT-TFA, 10,11-dehydro-NT-TFA, and DNT-TFA are in order of increasing retention times. Focusing upon the fragments $m/e = 217$, 219, and 232. Lower panel: Mass spectrum of synthetic 10,11-dehydro-NT-TFA.

VI. QUANTITATIVE MASS FRAGMENTOGRAPHY

Quantitative MF generally requires an internal standard. The idea behind the use of an internal standard is that it should compensate for errors introduced by factors which are otherwise difficult to control, such as variations in the yield of extraction and/or derivative formation, or losses due to absorption. From a chemical point of view, an ideal internal standard added at the beginning of a procedure should behave in a manner identical to the compound to be determined. Compounds labeled with stable isotopes fulfill all requirements. They differ only in mass from the compounds to be determined. No conventional system is capable of simultaneous determination of such substances. MF, however, has the unique ability to discriminate between such closely related compounds and to permit quantitative determination. The use of stable isotopes has been demonstrated by Gaffney, Hammar, Holmstedt, and McMahon (1971) and Borgå, Palmér, Sjöqvist, and Holmstedt (1973). It should be pointed out here that the difference in mass between the ions used for detection of the compound and the internal standard must be at least two mass units, preferably greater, to avoid interference of the two compounds due to the natural isotope content.

A quantitative method has been developed for the determination of nortriptyline in plasma, utilizing mass fragmentography (Borgå, Palmér, Linnarson, and Holmstedt, 1971). In this method a compound structurally related to nortriptyline was used as an internal standard. A metabolite, desmethylnortriptyline, could be analyzed in the same run. The compounds are first converted to their heptafluorobutyramide derivatives. Quantitative determinations are possible down to a few nanograms of nortriptyline per milliliter. Deuterium-labeled nortriptyline has also been used as an internal standard for quantitative analysis of nortriptyline in biological samples by Borgå et al. (1973). In addition to nortriptyline and other metabolites, 10-OH-NT has been determined and quantitated as DH_2NT-TFA in pharmacokinetic and genetic studies (Alexanderson, 1972, 1973).

A definite identification of acetylcholine in fresh rat brain was made possible for the first time by means of GC-MS and MF (Hammar, Hanin, Holmstedt, Kitz, Jenden, Karlén, 1968a). By means of MF it could also be concluded that the homologues propionylcholine and butyrylcholine were not present in significant amounts (< 1 ng/g brain). This meant that synthetic propionylcholine could be added to the rat brain homogenate as an internal standard, replacing the less suitable hexyltrimethylammonium bromide previously used.

Ion pair extraction with hexanitrodiphenylamine can be used for isolation of ACh in almost quantitative yield, not only in the nanomole range but also in picomole amounts. The isolated ion pairs can then be demethylated either chemically or by controlled pyrolysis before separation and quantitation (Karlén, Lundgren, Nordgren, and Holmstedt, 1973). The sensitivity of the two methods is about the same as that for bioassay. The limit of quantitation of ACh with these methods is about 1 nmole. When attempting to analyze acetylcholine in very low amounts the sensitivity of the hydrogen flame ionization detection is too low. To obtain sufficient sensitivity one has to use mass fragmentography. Deuterated ACh was used as an internal standard. In this way the sensitivity has been increased, making it possible to estimate samples of biological tissues containing about 10 pmoles of ACh.

Only a few examples of quantitative MF have been given in some detail above. Biogenically occurring compounds have been quantitated in addition to drugs, e.g. prostaglandins (Samuelsson, Hamberg, and Sweeley, 1970), 5-hydroxyindole-3-acetic acid (Bertilsson, Atkinson, Althaus, Härfast, Lindgren, and Holmstedt, 1972), indole-3-acetic acid (Bertilsson and Palmér, 1972), homovanillic acid (Sjöqvist and Änggård, 1972), and dopamine (Costa, Green, Koslow, LeFevre, Revuelta, and Wang, 1972).

VII. INTERCALIBRATION

The use of the GC-MS equipment is expensive. Whenever simpler and cheaper methods are available they should be used provided that they are equally accurate. It would seem that MF will be used on an increasing scale in the future for checking of other procedures. Examples of this are already available. The *in vitro* ^{3}H-acetylation method of NT was checked against MF (Borgå et al., 1971). Desmethylnortriptyline (DNT), which is a primary amine, is known to interfere in the ^{3}H-acetylation procedure. It was therefore expected that the latter method would give higher values of NT than those obtained by mass fragmentography. In 19 investigated patient plasma samples there was, in the majority of cases, a good agreement between the two methods. However, statistical analysis showed that the acetylation procedure had a tendency to give higher values than the mass fragmentographic one ($p < 0.05$). The plasma concentrations of DNT investigated in nine of the samples were of an order of magnitude to partly explain this discrepancy.

The quantitative measurement of p,p'-DDE in human plasma with mass fragmentography was recently performed to confirm the results obtained with the electron capture detector (Palmér and Kolmodin-Hedman, 1972). The mass fragmentographic method was used on 30 plasma samples where the concentration of p,p'-DDE was found to be in the 6 to 60 ng/ml range. Comparisons with the results obtained by electron capture detection gave a very good agreement between the two methods, $r = 0.99$. The specificity of this technique eliminated the uncertainty of compounds hidden under high p,p'-DDE peaks. (Fig. 4). In the future, techniques depending on fluorescence and immunoassay may be incalibrated with MF.

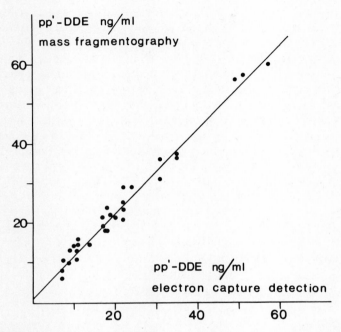

FIG. 4. Relationship between concentration of p,p'-DDE in plasma samples analyzed with electron capture and mass fragmentographic detection techniques.

VIII. CONCLUSIONS

Mass fragmentography is one of the most sensitive of all the gas chromatographic detection systems known. It can only be matched by an electron capture system for certain compounds. In general, MF is 100 to 1.000 times superior to FID and GC-MS. The sensitivity is obtained by selection of intense ions, wide optical slits, and a favorable signal-to-noise ratio which cannot be obtained in repetitive scanning. The possibility of utilizing certain fragments or ions for detection allows a unique means of selectivity, which can easily be changed in such a way that either a single compound or a family of related compounds can be recorded. By refocusing, "partial mass spectra" characteristic of the compounds may be obtained in spite of the fact that the amounts available are too small for the scanning of a complete mass spectrum.

The following criteria are used to characterize a compound by MF: the retention time of the compound, the presence of all the investigated mass numbers, and the characteristic ratio between their intensities. Although it is very selective and sensitive, the technique of MF may, in contrast to the electron capture detection, be applied to virtually any compound which can be gas chromatographed and has a suitable fragmentation pattern. The principle of mass fragmentography may allow simultaneous recordings of both parent drug and metabolites. MF is not restricted to identification purposes. It has already been applied to the quantitative analysis of drugs and endogenous compounds in biological samples such as blood or cerebrospinal fluid. It allows the ideal internal standards (i.e., compounds labeled with stable isotopes) to be used for quantitative determinations. Compounds labeled with stable isotopes can be given to man, and the metabolites are recognized because of specific ion clusters.

MF may be ideal for evaluation of the validity of simpler and less expensive methods. It has already been used with both magnetic and quadrupole mass spectrometers. It is to be anticipated that other techniques of mass spectrometry such as chemical ionization in the future will be combined with mass fragmentography.

ACKNOWLEDGMENTS

This work was supported by grants from the Swedish Medical Research Council 40Y-2375-05, 70E-3743-01; the Tri-Centennial Fund of the Bank of Sweden; the National Institute of Mental Health (MH 12007); the Wallenberg Foundation; the Swedish Environment Protection Board (2395-7-72); and by funds from Karolinska Institute.

REFERENCES

Alexanderson, B. (1972): On interindividual variability in plasma levels of nortriptyline and desmethylimipramine in man: A pharmacokinetic and genetic study. *Linköping's University Medical Dissertations*, No. 6.
Alexanderson, B. (1973): Predictions of steady-state plasma levels of nortriptyline from single oral dose kinetics, studied in twins. *European Journal of Clinical Pharmacology, in press.*
von Bahr, C. (1972): Binding and oxidation of amitriptyline and a series of its oxidized metabolites in liver microsomes from untreated and phenobarbital-treated rats. *Xenobiotica,* 2:293–306.
Bertilsson, L., Atkinson, Jr., A. J., Althaus, J. R., Härfast, Å., Lindgren, J.-E., and Holmstedt, B. (1972): Quantitative determination of 5-hydroxyindole-3-acetic acid in cerebrospinal fluid by gas chromatography-mass spectrometry. *Analytical Chemistry,* 44:1434–1438.

Bertilsson, L., and Palmér, L. (1972): Indole-3-acetic acid in human cerebrospinal fluid: Identification and quantification by mass fragmentography. *Science*, 177:74–76.
Bonelli, E. J. (1972): Gas chromatograph/mass spectrometer techniques for determination of interferences in pesticide analysis. *Analytical Chemistry*, 44:603–606.
Borgå, O., Palmér, L., Linnarson, A., and Holmstedt, B. (1971): Quantitative determination of nortriptyline and desmethylnortriptyline in human plasma by combined gas chromatography-mass spectrometry. *Analytical Letters*, 4:837–849.
Borgå, O., Palmér, L., Sjöqvist, F., and Holmstedt, B. (1973): Mass fragmentography used in quantitative analysis of drugs and endogenous compounds in biological fluids. In: *Symposium on the basis of drug therapy in man*. Fifth International Pharmacology Congress, San Francisco, vol. 3, C/9. S. Karger, Basel, *in press*.
Cattabeni, F., Koslow, S. H., and Costa, E. (1972): Gas chromatographic-mass spectrometric assay of four indole alkylamines of rat pineal. *Science*, 178:166–168.
Cho, A. K., Lindeke, B., Hodson, B. J., and Jenden, D. J. (1972): A gas chromatography/mass spectrometry assay for amphetamine in plasma. In: *Proceedings of the Fifth International Congress of Pharmacology*, p. 41. San Francisco, California.
Costa, E., Green, A. R., Koslow, S. H., LeFevre, J. F., Revuelta, A. V., and Wang, C. (1972): Dopamine and norepinephrine in noradrenergic axons: A study *in vivo* of their precursor product relationship by mass fragmentography and radiochemistry. *Pharmacological Reviews*, 24:167–190.
Elkin, K., Pierrou, L., Ahlborg, U. G., Holmstedt, B., and Lindgren, J.-E. (1973): Computer controlled mass fragmentography with digital signal processing. *Journal of Chromatography*, *in press*.
Gaffney, T. E., Hammar, C.-G., Holmstedt, B., and McMahon, R. E. (1971): Ion specific detection of internal standards labelled with stable isotopes. *Analytical Chemistry*, 43: 307–310.
Gohlke, R. S. (1962): Time-of-flight mass spectrometry: Application to capillary column gas chromatography. *Analytical Chemistry*, 34:1332–1333.
Hammar, C.-G. (1971): Mass fragmentography and elemental analysis by means of a new and combined multiple ion detector-peak matcher device. *Acta Pharmaceutica Suecica*, 8:129–152.
Hammar, C.-G. (1972): Qualitative and quantitative analyses of drugs in body fluids by means of mass fragmentography and a novel peak matching technique. In: *Proceedings of the International Symposium on Gas Chromatography Mass Spectrometry*, edited by A. Frigerio, pp. 1–18. Tamburini Editore, Milano.
Hammar, C.-G., Alexanderson, B., Holmstedt, B., and Sjöqvist, F. (1971): Gas chromatography-mass spectrometry of nortriptyline in body fluids of man. *Clinical Pharmacology and Therapeutics*, 12:496–505.
Hammar, C.-G., Hanin, I., Holmstedt, B., Kitz, R. J., Jenden, D. J., and Karlén, B. (1968a): Identification of acetylcholine in fresh rat brain by combined gas chromatography-mass spectrometry. *Nature*, 220:915.
Hammar, C.-G., and Hessling, R. (1971): Novel peak matching technique by means of a new and combined multiple ion detector-peak matcher device. *Analytical Chemistry*, 43:298–306.
Hammar, C.-G., Holmstedt, B., Lindgren, J.-E., and Tham, R. (1969): The combination of gas chromatography and mass spectrometry in the identification of drugs and metabolites. *Advances in Pharmacology and Chemotherapy*, 7:53–89.
Hammar, C.-G., Holmstedt, B., and Ryhage, R. (1968b): Mass fragmentography. Identification of chlorpromazine and its metabolites in human blood by a new method. *Analytical Biochemistry*, 25:532–548.
Henneberg, D. (1959): Ein kontinuierliches Verfahren zur massenspektrometrischen Bestimmung gaschromatographisch vorgetrennter Substanzgemische. *Zeitschrift für Analytische Chemie*, 170:365–366.
Henneberg, D. (1961): Eine Kombination von Gaschromatograph und Massenspektrometer zur Analyse organischer Stoffgemische. *Zeitschrift für Analytische Chemie*, 183:12–23.
Henneberg, D., and Schomburg, G. (1962): Mass spectrometric identification in capillary gas chromatography. *Gas Chromatography*, 1962:191–203.

Hites, R. A., and Biemann, K. (1970): Computer evaluation of continuously scanned mass spectra of gas chromatographic effluents. *Analytical Chemistry*, 42:855–860.

Holmstedt, B., and Linnarson, A. (1972): Chemistry and means of determination of hallucinogens and marihuana. In: *Drug Abuse — Proceedings of the International Conference*, edited by C. J. D. Zarafonetis, pp. 291–305. Lea & Febiger, Philadelphia.

Jenden, D. J., and Cho, A. K. (1972): Applications of integrated gas chromatography/mass spectrometry in pharmacology and toxicology. *Annual Review of Pharmacology*, 13:371–390.

Jenden, D. J., and Silverman, R. W. (1973): A multiple specific ion detector and analog data processor for a gas chromatograph/quadrupole mass spectrometer system. *Journal of Chromatography Science, in press.*

Karlén, B., Lundgren, G., Nordgren, I., and Holmstedt, B. (1973): Ion pair extraction in combination with gas phase analysis of acetylcholine. In: *Advances in Neuropsychopharmacology*, edited by Z. Votava. Avicenum, Prague, *in press.*

Knapp, D. R., and Gaffney, T. E. (1972): Commentary. Use of stable isotopes in pharmacology-clinical pharmacology. *Clinical Pharmacology and Therapeutics*, 13:307–316.

Knapp, D. R., Gaffney, T. E., and McMahon, T. E. (1972a): Use of stable isotope mixtures as a labeling technique in drug metabolism studies. *Biochemical Pharmacology*, 21:425–429.

Knapp, D. R., Gaffney, T. E., McMahon, R. E., and Kiplinger, G. (1972b): Studies of human urinary and biliary metabolites of nortriptyline with stable isotope labeling. *Journal of Pharmacology and Experimental Therapeutics*, 180:784–790.

Kolor, M. G. (1972): Flavor components. In: *Biochemical Application of Mass Spectrometry*, edited by G. R. Waller, Wiley-Interscience, New York.

Lindeman, L. P., and Annis, J. L. (1960): Use of a conventional mass spectrometer as a detector for gas chromatography. *Analytical Chemistry*, 32:1742–1749.

Palmér, L., and Kolmodin-Hedman, B. (1972): Improved quantitative gas chromatographic method for analysis of small quantities of chlorinated hydrocarbon pesticides in human plasma. *Journal of Chromatography*, 74:21–30.

Samuelsson, B., Hamberg, M., and Sweeley, C. C. (1970): Quantitative gas chromatography of prostaglandin E_1 at the nanogram level. *Analytical Biochemistry*, 38:301–304.

Sjöqvist, B., and Änggård, E. (1972): Gas chromatographic determination of homovanillic acid in human cerebrospinal fluid by electron capture detection and by mass fragmentography with a deuterated internal standard. *Analytical Chemistry*, 44:2297–2301.

Sweeley, C. C., Elliott, W. H., Fries, I., and Ryhage, R. (1966): Mass spectrometric determination of unresolved components in gas chromatographic effluents. *Analytical Chemistry*, 38:1549–1553.

Waller, G. R., Ed. (1972): *Biochemical Application of Mass Spectrometry*, Wiley-Interscience, New York.

Wiesendanger, H. U. D., and Tao, F. T. (1970): In: *Recent Development in Mass Spectroscopy*, edited by K. Ogata and T. Haikawa, pp. 290–295. University of Tokyo Press, Japan.

Advances in Biochemical Psychopharmacology, Vol. 7
Raven Press, New York © 1973

Chemical Ionization Mass Spectrometry

E. C. Horning, M. G. Horning, D. I. Carroll,
I. Dzidic, and R. N. Stillwell

Institute for Lipid Research, Baylor College of Medicine, Houston, Texas 77025

Until very recently, mass spectrometry was a field of interest only to chemists and physicists. The design of mass spectrometers was dictated largely by their analytical uses in hydrocarbon (petroleum) chemistry or by special requirements of physicists and physical chemists (who often built their own instruments from available components). This situation changed when Ryhage (1964) developed the "molecule separator" and a combined gas chromatograph-mass spectrometer (LKB-9000) was designed for use in biological and medical research studies. Since that time, the requirements for increased sensitivity of detection, for rapid or repetitive scanning, and for operation as part of an analytical system involving flowing gases have increasingly led to considerations of redesign of mass spectrometers. One of the issues under study, which is of considerable importance because of the uses of GC-MS (gas chromatograph-mass spectrometer) and GC-MS-COM (gas chromatograph-mass spectrometer-computer) analytical systems in pharmacology and toxicology, is the way in which ions are formed in the ion source of a mass spectrometer.

The electron impact ionization process, used in most mass spectrometers, involves the bombardment of vaporized compounds in the source with electrons generated from a heated filament. The efficiency of ionization is low, but the energy transfer when ionization occurs is high (the usual electron energy range is 20 to 70 eV). The initial step is formation of a positive ion ($M^{\pm}$) from the molecule, but the energy transfer is usually so great that fragmentation occurs, forming many ions. Cleavages at "weak" bonds occur to yield ions of expected structure, but rearrangements and eliminations of unsuspected nature often occur. The intensive study of fragmentation processes has provided much insight into the relationship between molecular structure and fragment ions, and mass spectrometry with electron impact (EI) conditions is highly useful in structural studies.

When the technique of mass spectrometry is used for quantitative purposes in GC-MS and GC-MS-COM systems, the formation of many ions from a single compound is almost always disadvantageous. The ionization process should be as efficient as possible (to increase sensitivity of detection) and should yield only a few instead of many fragment ions (to avoid interference in selective ion detection procedures). This is usually possible when chemical ionization (CI) processes are used for ion formation. The general term "chemical ionization" is used to describe ionization which occurs by an ion-molecule reaction. It is necessary, of course, to establish a condition which results in a primary ionization step. Since gas-phase ion-molecule reactions occur with great speed, a complex series of reactions may occur (depending on the conditions in the ion source) yielding one or a few ions from the neutral molecules under study. Ion-molecule reactions usually involve comparatively low energies.

Two kinds of chemical ionization reaction conditions are of interest for their applications in pharmacology and toxicology. Most mass spectrometers can be modified for chemical ionization with gases under pressures of 1 to 2 torr without changing the source location or sample inlet arrangements, although modification of the vacuum system may be needed. It is also possible to carry out ion-molecule reactions at atmospheric pressure in an "external" source; a novel mass spectrometer with very high sensitivity in detection has been devised utilizing this design. The following sections describe a number of chemical ionization methods and their applications.

BASIC PRINCIPLES

An ion and a neutral molecule, present together in the usual internal or in an external source, will react if an exothermic reaction can occur. This may be expressed in terms of ionization potentials, or in the usual form of a chemical reaction. Virtually no activation energy is required. The collision may result in charge transfer alone, but addition of an ion is also a common process. If the resulting structure is stable, no further reaction occurs. If not, a simple cleavage will occur. This is the route followed in proton transfer reactions and in proton abstraction. When ions formed by charge transfer or proton transfer are not sufficiently stable, functional group elimination may occur. When this happens, the mass spectrum usually shows M^+ or MH^+ and typical ions due to loss of neutral fragments such as trimethyl-silanol. The formation of many ion radicals is usually due to high energy processes which are not usually seen under CI conditions.

In ordinary mass spectrometers, the CI source is usually operated with methane, isobutane, or nitrogen-ammonia; other reagents that have been used include tetramethylsilane and nitric oxide. The pressure in the source is usually approximately 1 torr (sometimes referred to as a high-pressure condition, in contrast to the usual EI condition where the source pressure is only slightly above the pressure in the mass analyzer), and the primary ionization step takes place through electron impact (the electrons are generated from a filament, as usual) with the reagent gas. Most of the original work was with methane, and later with isobutane; the background and early research studies are described in a review by Field (1968).

A. Proton Transfer Reactions

$$M + CH_5^+ \rightarrow MH^+ + CH_4 \tag{1}$$
$$M + C_4H_9^+ \rightarrow MH^+ + C_4H_8 \tag{2}$$
$$M + C_6H_6^{+\cdot} \rightarrow MH^+ + C_6H_5 \tag{3}$$
$$M + NH_4^+ \rightarrow MH^+ + NH_3 \tag{4}$$
$$M + H_3O^+ \rightarrow MH^+ + H_2O \tag{5}$$

Proton transfer is perhaps the most widely used chemical ionization reaction. If M is a strong base in the gas phase (a basic drug, for example), MH^+ will be formed by reaction with proton donors derived from water or ammonia, from ion radicals such as those derived from benzene, or from reaction with the very powerful proton donor CH_5^+ from methane. If M is a weak base (weaker than ammonia or water, for example), reaction with ions from methane or isobutane may be required to form MH^+. The most powerful proton donor known in gas phase reactions is CH_5^+.

If the ion MH^+ is stable, the mass spectrum may show only this ion (in addition to the reagent ions). In some instances, however, additional ions may be present because of fragmentation, or because stable addition products are formed.

B. Addition Reactions

$$M + NH_4^+ \rightarrow MHNH_3^+ \tag{6}$$
$$M + H_3O^+ \rightarrow MHH_2O^+ \tag{7}$$
$$M + C_2H_5^+ \rightarrow MC_2H_5^+ \tag{8}$$
$$M + C_3H_7^+ \rightarrow MC_3H_7^+ \tag{9}$$
$$M + MH^+ \rightarrow MHM^+ \tag{10}$$
$$M + M^{+\cdot} \rightarrow MM^{+\cdot} \tag{11}$$
$$H_2O + H_3O^+ \rightarrow H_2OHH_2O^+ \tag{12}$$

In many instances a polar molecule, M, will react with a positively charged ion to yield an addition product. For example, a basic drug may show ions MH^+ (reaction 4) and MNH_4^+ (reaction 6) when ammonia is used as a reagent gas. The ion ratio will depend upon the structure of M and the reaction conditions (temperature, pressure, concentration). Reactions 8 and 9 are usually observed whenever methane is used as a reagent gas, since ethyl and isopropyl ions are present at the same time. The CI mass spectrum of a barbiturate in methane, for example, will contain MH^+ (from reaction 1), $MC_2H_5^+$ (from reaction 8), and $MC_3H_7^+$ (from reaction 9). Water is rarely used for CI purposes in conventional mass spectrometers, but the formation of cluster ions from the reagent (reaction 12, continuing to form larger ions) is a familiar occurrence in reactions carried out at atmospheric pressure. A kind of dimerization reaction, in which M may react with either M^+ or MH^+, may also occur. An example of reaction 11 is the formation of $C_{12}H_{12}^+$ from $C_6H_6^+$ when benzene is a reagent at atmospheric pressure. The possible formation of MHM^+ (reaction 10) should be considered whenever MH^+ is the major reaction product; the occurrence of this reaction depends both on the structure of M and on the reaction conditions.

C. Proton Transfer with Elimination or Cleavage

$$M + CH_5^+ \rightarrow (M\text{-}H)^+ + H_2 + CH_4 \tag{13}$$

$$M + CH_5^+ \rightarrow (M\text{-}H_2O)^+ + H_2O + CH_4 \tag{14}$$

$$M + CH_5^+ \rightarrow (MH\text{-}90)^+ + (CH_3)_3SiOH + CH_4 \tag{15}$$

The ion CH_5^+ is a sufficiently strong proton donor to react with hydrocarbons. If M is an alkane, for example, reaction 13 will occur; the initial adduct is not stable and the reaction is one of proton abstraction. A similar reaction may occur for alicyclic compounds; steroids, for example, often show $(M\text{-}1)^+$ or related ions when methane is the reagent gas. This reaction does not occur with ammonia, water, or benzene. An initially formed MH^+ will often undergo cleavage when a powerful proton donor is used. For example, if M is a monofunctional alcohol and methane is the reagent gas, reaction 14 will occur. If M contains a trimethylsilyl ether group, the usual reaction is protonation followed by loss of trimethylsilanol (reaction 15). If two or three trimethylsilyl ether groups are present, the ions which are formed are usually MH^+, $(MH\text{-}90)^+$, $(MH\text{-}180)^+$, and additional ions formed by repetition of the elimination process.

Although these reactions are generally believed to be undesirable, they are in fact valuable in some applications. For example, conditions are known under which all hydroxyl groups, even those most highly hindered with respect to chemical reaction, can be converted to trimethylsilyl ethers (Thenot and Horning, 1972); the number of hydroxyl groups can be counted easily by the elimination steps and by the use of deuterated reagents.

D. Charge Transfer, with or without Fragmentation

$$M + C_6H_6^{+\cdot} \rightarrow M^{+\cdot} + C_6H_6 \tag{16}$$
$$M + C_6H_6^{+\cdot} \rightarrow (M\text{-}90)^{+\cdot} + (CH_3)_3SiOH + C_6H_6 \tag{17}$$

Most CI work has been carried out with methane, isobutane, or ammonia, with emphasis in drug studies on ions of the MH^+ type. It is also possible to ionize compounds by charge transfer, if an appropriate reagent gas of higher ionization potential is used. For example, trimethylsilyl ethers have low ionization potentials. Steroid trimethylsilyl ethers can be ionized in the presence of benzene (reaction 16), but this is followed by loss of trimethylsilanol (reaction 17). The relative proportions of ions which are formed depend upon the structure of the parent compound.

E. Proton Abstraction with Negative Ion Formation

$$M + Cl^- \rightarrow (M\text{-}H)^- + HCl \tag{18}$$

Acidic substances, in the presence of ions of basic character, will lose a proton to form $(M\text{-}H)^-$ (reaction 18). Reactions of this kind are not observed in mass spectrometers unless provision has been made for operation in a negative ion mode. The reaction probably takes place through an intermediate MCl^-.

F. Primary or Initial Ionization Reactions

$$CH_4 \rightarrow CH_4^{+\cdot} + CH_3^+ + CH_2^{+\cdot} \tag{19}$$
$$CH_4 \rightarrow CH_5^+ + C_2H_5^+ + C_3H_7^+ \tag{20}$$

When methane is ionized under EI conditions, the product ions are largely those expected from a high-energy, low-pressure ionization condition (reaction 19). If the concentration of methane is increased to 0.5 to 1 torr, a series of secondary reactions occurs and the ion products are those shown in reaction 20. The ion formed in highest yield (47%) is CH_5^+, but ethyl and isopropyl ions are also present (41% and 6%, respectively). Since polar molecules will add ethyl and isopropyl ions as well as protons, the methane CI spectra of bases will normally show MH^+, $(M + 29)^+$, and $(M + 41)^+$. A similar but less marked effect occurs when isobutane is used as the reagent gas; the major ion product is MH^+, but ions at $(M + 57)^+$ and $(M + 39)^+$ may be seen. For this reason, and because the energy transfer is lower, isobutane is preferred as a reagent gas in many applications involving bases.

G. Reactions in Nitrogen in an External Source

$$N_2 + e \rightarrow N_2^{+\cdot} + 2e \tag{21}$$
$$N_2^{+\cdot} + 2N_2 \rightarrow N_4^{+\cdot} + N_2 \tag{22}$$
$$N_4^{+\cdot} + H_2O \rightarrow H_2O^{+\cdot} + 2N_2 \tag{23}$$
$$H_2O^{+\cdot} + H_2O \rightarrow H_3O^+ + OH \tag{24}$$
$$N_2 + H_3O^+ + H_2O \rightarrow H_2OHH_2O^+ + N_2 \tag{25}$$
$$N_4^{+\cdot} + C_6H_6 \rightarrow C_6H_6^{+\cdot} + 2N_2 \tag{26}$$
$$H_2O^{+\cdot} + C_6H_6 \rightarrow C_6H_6^{+\cdot} + H_2O \tag{27}$$
$$C_6H_6^{+\cdot} + C_6H_6 \rightarrow C_{12}H_{12}^{+\cdot} \tag{28}$$

When nitrogen containing a small amount of water is subjected to ionizing conditions, a complex series of reactions occurs leading to the formation of ion clusters of protonated water. The initial reaction is with nitrogen, and the products are formed through charge transfer and proton transfer reactions. These ion-molecule reactions, studied by Good, Durden, and Kebarle (1970), will occur in a ^{63}Ni reaction cell. If benzene vapor is present, the water ions and nitrogen ions react with the organic compound to form benzene radical ions (reactions 26 and 27) through charge transfer. A dimer ion is also formed (reaction 28). These ions can participate in proton transfer or charge transfer reactions with other substances with a lower ionization potential.

II. SPECTRA

Figure 1 shows the EI (70 eV) mass spectrum of the N,N-dimethyl derivative of phenobarbital. When a CI spectrum of a barbiturate derivative is obtained in methane, the chief ion is MH^+, and ions corresponding to $(M + 29)^+$ and $(M + 41)^+$ are also present (Fig. 2). Cleavage occurs to a very small degree, in contrast to the effect of the high-energy fragmentation process illustrated in Fig. 1. With isobutane as the carrier and reagent gas, cleavage does not occur and the ion product is MH^+ (Fig. 3), together with some $M^{+\cdot}$ formed by charge transfer. Amines are protonated. Figures 4 and 5 show CI (methane) mass spectra for nicotine and methadone. The chief ion product in each instance is MH^+; ions at $(M + 29)^+$ and $(M + 41)^+$ are also present as minor reaction products. Results of this kind were obtained in all early studies of methane chemical ionization spectra. Fales and his associates (Milne, Fales, and Axenrod, 1971) emphasized the simplicity of the mass spectra, and the formation of MH^+ as the principal ion from many drugs. Cleavage will occur, however, for many nonbasic substances

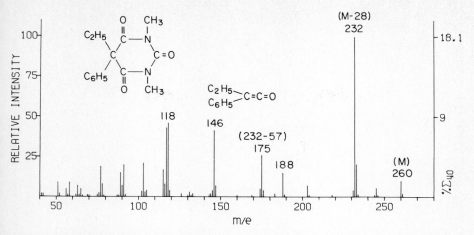

FIG. 1. Electron impact (EI-70 eV) mass spectrum of the N,N-dimethyl derivative of pheno-barbital. Many fragment ions are present. Instrument: LKB 9000.

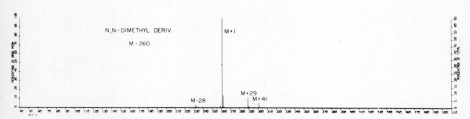

FIG. 2. Methane chemical ionization (CI) spectrum of the N,N-dimethyl derivative of phenobarbital. The principal ion is MH$^+$, and MC$_2$H$_5$$^+$ and MC$_3$H$_7$$^+$ are also present. A very small yield of (M − 28)$^+$ was observed; this is the chief ion obtained under EI conditions (see Fig. 1). The simplicity of the ion profile should be compared with that shown in Fig. 1. Instrument: Finnigan 1015.

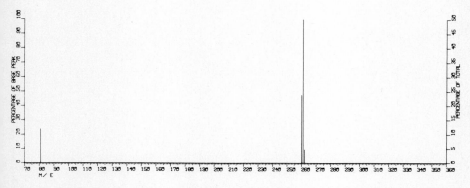

FIG. 3. Isobutane chemical ionization (CI) mass spectrum of the N,N-dimethyl derivative of phenobarbital. The principal ion is MH$^+$, along with M$^{\cdot}$ formed by charge transfer. There are no cleavage products. This mass spectrum should be compared with those in Figs. 1 and 2. Instrument: Finnigan 1015.

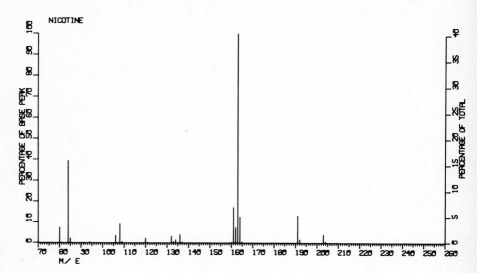

FIG. 4. Methane chemical ionization (CI) mass spectrum of nicotine. The principal ion is MH⁺; ions at M + 29⁺ and M + 41⁺ are also present, together with a few fragment ions. Instrument: Finnigan 1015.

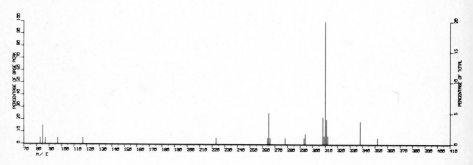

FIG. 5. Methane chemical ionization (CI) mass spectrum of methadone. The principal ion is MH⁺; ions at M + 29⁺ and M + 41⁺ are also present, together with a few fragment ions. Instrument: Finnigan 1015.

when methane is used as a carrier and reagent gas. For long chain compounds and steroids, ions at $(M - 1)^+$ or cleavage products of $(M - 1)^+$ are usually seen. Figure 6 shows the CI (methane) spectrum of ethyl heptadecanoate; the major ion products are $(M + 1)^+$ and $(M - 1)^+$. Figure 7 shows the CI (methane) spectrum of the MO − TMS derivative of androsterone. The major ion products correspond to protonation and to functional group cleavage. Additional CI spectra of steroids were described by Vanden-Heuvel (Smith and VandenHeuvel, 1972).

When ammonia is used as a reagent, basic drugs are protonated. Ions corresponding to $(M + NH_4)^+$ are sometimes observed, but cleavage products are rarely seen.

It should be emphasized that high-energy (EI) processes lead to fragmentation products which include odd electron (radical) ions, as well as even electron ions. Cleavages which occur under CI conditions do not usually yield radical ions; the functional group cleavage reactions which have been observed correspond to the elimination of water, methanol, trimethylsilanol, or other molecular entities. For example, a steroid TMS derivative will usually show $M^{\ddot{+}}$ and $(M - 90)^{\ddot{+}}$ ions under EI conditions; the corresponding ions in a CI reaction will be MH^+ and $(MH - 90)^+$ under conditions of proton transfer. If charge transfer occurs, however, ions corresponding to $M^{\ddot{+}}$ and $(M - 90)^{\ddot{+}}$ will be seen.

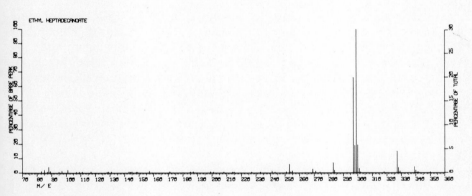

FIG. 6. Methane chemical ionization (CI) spectrum of ethyl heptadecanoate. The principal ions are MH^+ and $(M-H)^+$; $M + 29^+$ and $M + 41^+$ are also present. Instrument: Finnigan 1015.

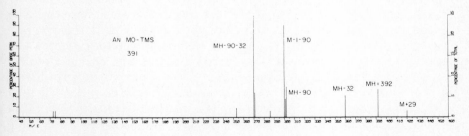

FIG. 7. Methane chemical ionization (CI) spectrum of the methoximetrimethylsilyl (MO-TMS) derivative of androsterone. The principal ions are formed by cleavage reactions, but there is an appreciable yield of MH^+. Instrument: Finnigan 1015.

III. APPLICATIONS

Qualitative and quantitative information about drugs and their metabolites in body fluids can be obtained through the use of GC-MS or GC-MS-COM systems. Three types of procedures have been used. The oldest method, which is still widely used, is to obtain EI mass spectra by manual triggering whenever a peak of interest is noted as the elution proceeds (the total ion current monitor is used for mass detection purposes). No great advantage is gained by using CI ionization techniques under these circumstances; the objective is usually that of identifying a drug or drug metabolite, and an EI mass spectrum may be more valuable than a CI spectrum for this purpose. A second method is to employ repetitive scanning, followed by computer-based programmed analysis of the data, to detect specific compounds, or more correctly, the occurrence of specific ions known or believed to be characteristic of the compounds under study. This was called "mass chromatography" by Biemann (Hites and Biemann, 1970). The use of CI techniques is frequently advantageous; for example, if a basic drug is under study, the ion characteristic of the drug would be MH^+. Very little interference normally occurs, and the data can be retained if desired for analysis or reexamination at a later time. The major disadvantage lies in the fact that high sensitivity of detection is not usually possible. The scanning process requires equal time for estimating ion intensities at each amu value over a large mass range. Further, the amount of stored data from a single GC-MS-COM run may be great.

The third procedure is to focus upon one or a few ions, through use in magnetic instruments of circuits which control the accelerating voltage, so that virtually continuous observation of the occurrence of these ions can be achieved. This is generally called selective ion monitoring; the term introduced by Holmstedt, "mass fragmentography," also describes the process (Hammar, Holmstedt, and Ryhage, 1968). The chief advantage of this technique, when compared with repetitive scanning, is that maximum sensitivity of detection can be achieved, and this is why it is often preferred over programmed analysis methods. The precise method of monitoring varies with local equipment and preferences; a hardware device with analogue recording can be employed, or the ion-monitoring process can be computer driven. Few or many ions may be monitored, but the limits of detection will alter if a large number of ions are monitored. It is also possible to carry out selective ion monitoring by varying the field strength, but this is rarely done with magnetic instruments.

At present the chief use of these methods is in the determination of serum or tissue concentrations of drugs or other compounds which have relatively high biological activity; the practical limit of detection at this time is approximately 5 ng/ml of serum for most drugs that have been studied. Other chapters in this volume contain information about specific applications, such as those of Costa and his colleagues (Cattabeni, Koslow, and Costa, 1972).

Figures 8 and 9 show a different way of using selective ion monitoring

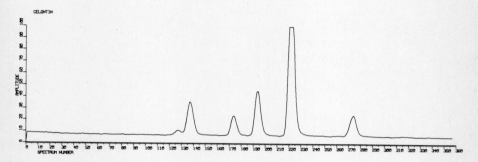

FIG. 8. Detection of MH⁺ ions corresponding to the TMS derivatives of monohydroxy metabolites of Celontin ®. Six isomers are present. Conditions: methane chemical ionization. Finnigan 1015 mass spectrometer.

techniques. It is relatively easy to obtain metabolic profiles for specific groups of compounds when CI methods are used. For example, hydroxylation is a general route of drug metabolism for many drugs; the molecular weight of the TMS derivative of a monohydroxy compound can be calculated, and if the parent drug forms an MH⁺ ion, it may be expected that protonated ions will also be formed for the monohydroxy metabolites. In the example in Fig. 8, each peak corresponds to the same molecular formula — that of the protonated TMS derivative of monohydroxy Celontin ® (N,2-dimethyl-2-phenylsuccinimide). Six isomers are formed. By repeating the run, with monitoring for the methoxy derivative of a dihydroxy Celontin®, it was found that two isomers were formed (Fig. 9). If the detection of aromatic hydroxylation products is desired, the mixture of metabolites is treated with an excess of diazomethane for several minutes to methylate phenolic groups. A search can then be made for the appropriate compounds.

The success of this approach for the rapid detection of groups of drug metabolites lies in the fact that CI spectra can usually be predicted, and very little interference from other compounds is likely to occur. It is possible to look directly for predicted compounds resulting from metabolic pathways known to occur under the circumstances of the study.

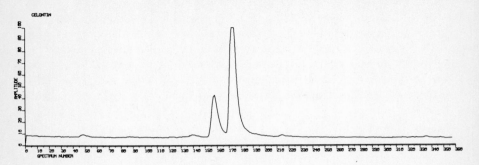

FIG. 9. Detection of MH⁺ ions corresponding to methyl ethers (aromatic) derived from dihydroxy metabolites of Celontin®. Two isomers are present. Conditions: methane chemical ionization, Finnigan 1015 mass spectrometer.

IV. A NEW PICOGRAM DETECTION SYSTEM

The designs of mass spectrometers used in GC-MS and GC-MS-COM systems are based on considerations which no longer apply. High sensitivity in detection was not usually a requirement in the past; now, it may be of paramount importance. The function of these analytical systems is to detect and quantify molecular entities in a flowing gas stream, but the use of internal sources and slow speed pumping assemblies is based on a different concept of function. High-energy ionization processes are employed largely because some of the major analytical applications in the past were in hydrocarbon (petroleum) chemistry. Most compounds of biologic interest can be ionized by comparatively low-energy processes; this is preferred when detection and quantification are primary aims, rather than the seeking of structural information.

A novel type of mass spectrometer with an external source operated at atmospheric pressure and with a ⁶³Ni source of electrons rather than a filament may prove to be a valuable detection system in biomedical studies. The external source is a small reaction chamber containing a ⁶³Ni foil; the carrier-gas (nitrogen) flow through the chamber may be from a gas chromatograph or from a simple heated injection device. A jet of gas enters the mass analyzer region through a 25-μ aperture. In effect, the mass spectrometer acts as a detector engaged in continuous sampling of the gas mixture in the reaction chamber. The quadrupole mass analyzer is of standard design (Finnigan) but arranged for operation in either positive or negative ion mode. Pulse-counting circuitry is used for ion detection; a small computer (PDP 8/E) with a laboratory interface is used for acquisition and analysis of data.

A complex series of ion-molecule reactions occurs when nitrogen containing a little water is used as a carrier-gas stream (the water is not added; it is not possible to remove water completely when ordinary gases and ordinary materials are used for tubing and reaction chamber construction) (reactions 21–25). When organic solvents are swept through the reaction chamber, additional ionization reactions occur. Methanol and ethanol form ion clusters analogous to those formed from water. Benzene yields $C_6H_6^+$ and $C_{12}H_{12}^+$ ions (reactions 26–28).

Many organic compounds are ionized when introduced into the reaction chamber. Basic drugs are protonated to yield MH^+; other compounds with basic properties in the gas phase also yield MH^+. Figure 10 shows a scanned mass spectrum obtained for a solution containing cocaine and methadone; benzene or chloroform may be used as solvents.

Very high sensitivity in detection has been attained. Figure 11 shows the records obtained by selective ion monitoring for MH^+ derived from 2,6-dimethyl-γ-pyrone. This compound is not highly adsorbed on glass or metal surfaces, and it is ionized in benzene vapor in the reaction chamber through protonation. The shape of the peaks is due to the injection process; a chromatographic column was not used. The linearity of response with respect to mass is shown in Fig. 12. Samples of approximately 5 pg can be

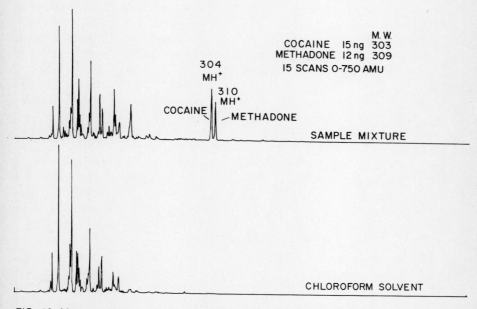

FIG. 10. Mass spectra of cocaine and methadone in chloroform. The ion products correspond to MH^+. The solvent ions are due to ethanol in the chloroform. The scan was a result of signal averaging for 15 scans from 0 to 750 amu. Instrument: ILR.

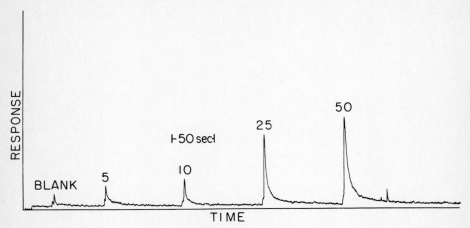

FIG. 11. Successive responses to injection of samples of 2,6-dimethyl-γ-pyrone in benzene. Single-ion monitoring mode was used for MH⁺. A 5-pg sample was easily detected;
the blank is believed to represent a small amount of the pyrone flushed from the injector-
reactor region by solvent. The peak shape is due to the direct method of injection. A
column was not used. Instrument: ILR.

detected, and the response is linear up to 10 ng. Figure 13 shows the results
obtained in a scanning (0 to 750 amu) mode. A sample of 25 pg gave a detectable ion.

In general, the ionization reactions found to occur in this type of source

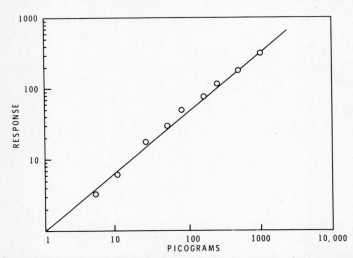

FIG. 12. Relationship of response (area under peaks) to mass for 2,6-dimethyl-γ-pyrone
injected in benzene solution, as shown in Fig. 11. The response is linear for amounts from
5 pg to 1 ng. Instrument: ILR.

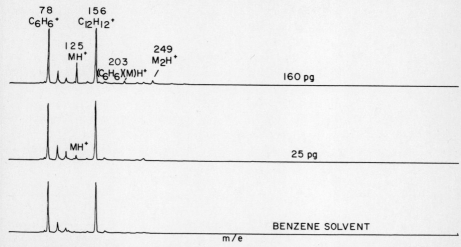

FIG. 13. Scanned mass spectrum (0 to 750 amu) for 2,6-dimethyl-γ-pyrone injected in benzene solution. The MH⁺ ion is detectable for a sample of 25 pg. Instrument: ILR.

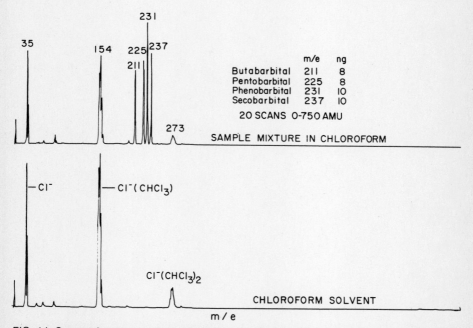

FIG. 14. Scanned mass spectrum (0 to 750 amu, negative ion mode) for four barbiturates injected in chloroform solution. The scan was signal averaged over 20 scans. The drugs form $(M - 1)^-$ ions due to ionization by Cl^- with proton removal. The negative ion mass spectrum of chloroform is shown in the lower chart. Instrument: ILR.

resemble those occurring in conventional sources under relatively low energy conditions. The ionization potential of benzene, for example, is estimated at 9.5 to 10.5 eV. Most silyl derivatives are ionized in the source in the presence of benzene through charge transfer reactions. Compounds with a higher ionization potential than benzene would not be ionized in the presence of a high concentration of gaseous benzene.

Negative ions are formed through electron capture or by ionization. When a chloroform solution of barbiturates was injected, all drug molecules underwent ionization through reaction with chloride ion to form $(M - 1)^-$. This result is shown in Fig. 14. Figure 15 shows the detection of phenobarbital and its major metabolites in rat urine; derivative formation was not employed, and a chromatographic column was not used.

Three possible ways of using this detection system are under investigation. These are (1) in the analysis of samples which do not require chromatography, (2) as part of an LC-MS-COM system, with some or all of the effluent stream vaporized through the source, and (3) as part of a GC-MS-COM system with columns of 100,000 or higher theoretical plate efficiency

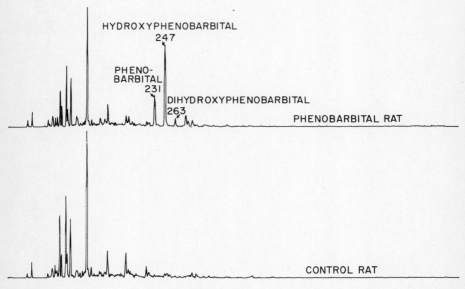

FIG. 15. Scanned mass spectrum (0 to 750 amu, negative ion mode) for phenobarbital and its major metabolites injected in methanol solution. The drug and metabolite mixture was obtained by extraction of underivatized rat urine collected in a 24- to 48-hr period after drug administration. This result should be compared with the chart shown in Fig. 14. Instrument: ILR.

made according to a new process (German and Horning, 1973; German, Pfaffenberger, Thenot, Horning, and Horning, 1973). The primary objective is to obtain very high sensitivity and specificity in detection for compounds of biologic interest. The external source concept could, of course, be used with other types of ionization conditions, but at present chemical ionization reactions appear to be suitable for many applications.

ACKNOWLEDGMENT

This work was aided by grants GM-13901 and GM-16216 of the National Institute of General Medical Sciences, grant HE-05435 of the National Heart and Lung Institute, grant Q-125 of the Robert A. Welch Foundation, and contract NIH 69-2161 of the National Institute of General Medical Sciences.

REFERENCES

Cattabeni, F., Koslow, S. H., and Costa, E. (1972): Gas chromatographic-mass spectrometric assay of four indole alkylamines of rat pineal. *Science,* 178:166–168.
Field, F. H. (1968): Chemical ionization mass spectrometry. *Accounts of Chemical Research,* 1:42–49.
German, A. L., and Horning, E. C. (1973): Thermostable open tube capillary columns for the high resolution gas chromatography of human urinary steroids. *Journal of Chromatographic Science,* 11:76.
German, A. L., Pfaffenberger, C. D., Thenot, J-P., Horning, M. G., and Horning, E. C. (1973): High resolution gas chromatography with thermostable glass capillary columns. *Analytical Chemistry, in press*
Good, A., Durden, D. A., and Kebarle, P. (1970): Ion-molecule reactions in pure nitrogen and nitrogen containing traces of water at total pressures 0.5–4 torr. Kinetics of clustering reactions forming $H^+(H_2O)_n$. *Journal of Chemical Physics,* 52:212–221.
Hammar, C-G., Holmstedt, B., and Ryhage, R. (1968): Mass fragmentography identification of chlorpromazine and its metabolites in human blood by a new method. *Analytical Biochemistry,* 25:532–548.
Hites, R. A., and Biemann, K. (1970): Computer evaluation of continuously scanned mass spectra of gas chromatographic effluents. *Analytical Chemistry,* 42:855–860.
Milne, G. W. A., Fales, H. M., and Axenrod, T. (1971): Identification of dangerous drugs by isobutane chemical ionization mass spectrometry. *Analytical Chemistry,* 43:1815–1820.
Ryhage, R. (1964): Use of a mass spectrometer as a detector and analyser for effluents emerging from high temperature gas-liquid chromatography columns. *Analytical Chemistry,* 36:759–762.
Smith, J. L., and VandenHeuvel, W. J. A. (1972): Chemical ionization mass spectrometry of steroids. *Analytical Letters,* 5:51–58.
Thenot, J.-P., and Horning, E. C. (1972): MO-TMS derivatives of human urinary steroids for GC and GC-MS studies. *Analytical Letters,* 5:21–33.

Advances in Biochemical Psychopharmacology, Vol. 7
Raven Press, New York © 1973

Analysis of Pineal and Brain Indole Alkylamines by Gas Chromatography – Mass Spectrometry

Stephen H. Koslow and A. Richard Green

Laboratory of Preclinical Pharmacology, National Institute of Mental Health, St. Elizabeths Hospital, Washington, D.C. 20032

Present quantitative methods for the measurement of the putative neuro-transmitters are neither specific nor sensitive enough to study the precise neurochemical role of these compounds at specific central nervous system nuclei and cell groupings. Although the use of the histofluorescent technique of Falck, Hillarp, Thieme, and Torp (1962) has greatly extended our knowledge of the neurotransmitters at the cellular level, these measurements are only semiquantitative, and they are not specific. We have applied the combined instrumentation of gas chromatography (GC) and mass spectrometry (MS) to the quantitative measurement of both catecholamines (Koslow, Cattabeni, and Costa, 1972) and indolealkylamines (Cattabeni, Koslow, and Costa, 1972). The combination of GC with MS allows for the analysis of impure samples, with sensitivity in the order of 10^{-7} mole. The specialized technique of mass fragmentography increases the sensitivity 10^7 times, thus extending the lower limit of sensitivity to 10^{-14} to 10^{-15} mole. In addition to having extreme sensitivity the variation of multiple-ion detection (MID) gives a high degree of specificity to the assay.

This chapter will deal solely with the measurement of the indolealkyl-amines. The prime indole in the central nervous system (CNS) is serotonin (S), a putative neurotransmitter. Other important indoles are 5-methoxy-tryptamine (5MT), N-acetylserotonin (NAS), and melatonin (M) (Fig. 1). These three compounds have been reported to be localized only in the pineal gland.

I. MATERIAL AND METHODS

In order to analyze the indoles by GC-MS it was necessary to increase their volatility. This was done by acylating the hydroxy and 1° and 2° amine groups with pentafluoropropionic anhydride (PFPA). For serotonin

SEROTONIN
5-HYDROXYTRYPTAMINE

5-METHOXYTRYPTAMINE

N-ACETYLSEROTONIN

MELATONIN

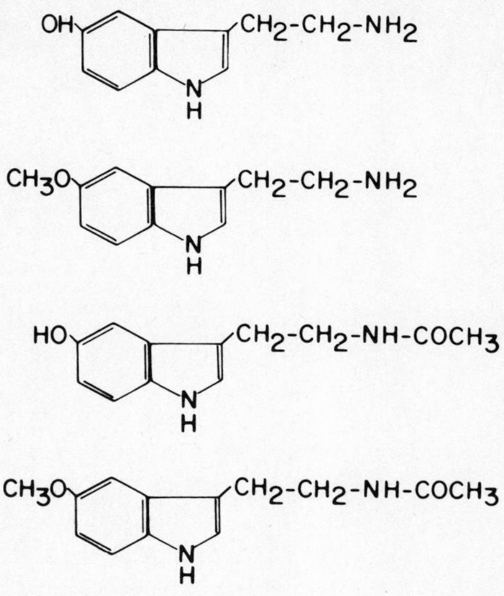

FIG. 1. Structural formulas of the four indolealkylamines assayed.

(Fig. 2) and 5-methoxytryptamine this reaction is simply addition of penta-fluoropropional (PFP) groups. On the other hand, when an indole containing an N-acetyl moiety is reacted with PFPA, there is in addition to the aforementioned reactions a cyclization of the side chain resulting in a structure which resembles a β-carboline (Fig. 2) (Koslow, Cattabeni, and Costa, 1973). Since two types of reactions occur, it was necessary to use two internal standards: α-methylserotonin (αMS) for S and 5MT, and N-acetyltryptamine (NAT) for NAS and M (Fig. 3). All six of these compounds when reacted with 100 μliter PFPA in 20 μl ethylacetate reach a steady state after 3 hr at 60°C. The acylated derivatives are stable for 24 hr both in the presence of PFPA or after the excess PFPA has been removed and the derivatives dissolved in ethylacetate.

All analyses were done on an LKB 9000 GC-MS with the following conditions: gas chromatography column 12 ft, (i.d., 2 mm) packed with 3% OV-17 on Gas Chrom Q; flash heater 230°C, oven 200°C, separator 250°C, ion source 290°C, trap current 60 μA, accelerating voltage 3.7 kV. For quantitation, the ion density of the base peak of the indole (the most abundant peak, 100%) is recorded as the compound is eluted from the GC (Table 1). A typical mass fragmentographic analysis of the six indole derivatives prepared from pure compounds is shown in Fig. 4. Standards are run daily in order to obtain calibration curves. A constant amount of

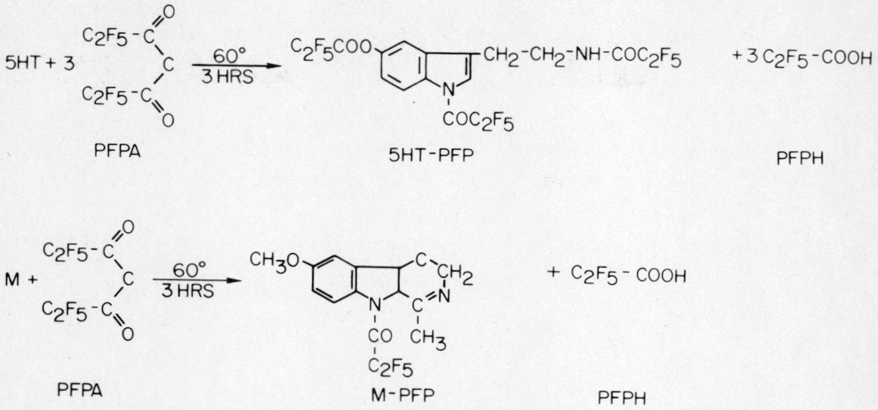

FIG. 2. Two main types of reactions occur when the indoles are reacted with pentafluoro-propionic anhydride (PFPA). For serotonin (5HT) and other primary amines, the available OH, NH, and NH_2 groups are acylated, resulting in the addition of three perfluorinated (PFP) groups. For melatonin and other N-acetylindoles, the reaction with PFPA results in the splitting out of a H_2O molecule and the formation of a tricyclic compound (M-PFP). In both cases there is a stochiometric formation of pentafluoropropionic acid (PFPH).

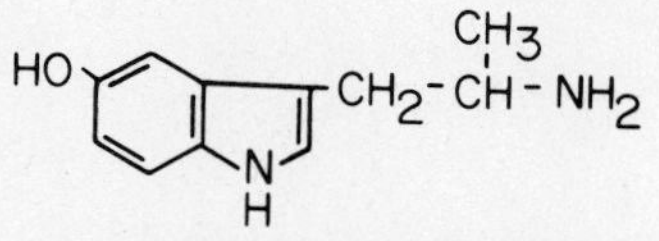

α-METHYL SEROTONIN

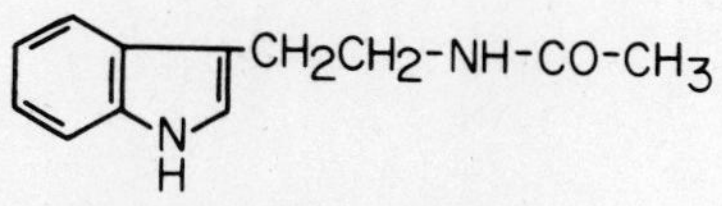

N-ACETYLTRYPTAMINE

FIG. 3. Two internal standards used for quantitation. α-Methylserotonin serves as internal standard for serotonin and 5-methoxytryptamine. N-acetyltryptamine is the internal standard for N-acetylserotonin and melatonin.

internal standards (40 pmole) is added to varying concentrations of the indoles (1 to 600 pmole) and analyzed by mass fragmentography. The ratio formed by dividing the peak height of the indole by the peak height of the appropriate internal standard is plotted against the known concentration of

TABLE 1. *Gas chromatographic (GC) retention times and positive ions monitored for quantitation of the 5-substituted indolealkylamine pentafluoropropionyl (PFP) derivatives. m/e, mass-to-charge. Relative intensity (%) refers to the relative intensity of the fragment compared to the most abundant (base) peak which is 100%.*

INDOLE PFP DERIVATIVE	GC TIME (MIN)	POSITIVE ION RECORDED FOR QUANTITATION	m/e	RELATIVE INTENSITY (%)
α-METHYLSEROTONIN	2.7		465	36
SEROTONIN	3.4		451	100
N–ACETYLSEROTONIN	4.5		492	100
N–ACETYLTRYPTAMINE	5.8		330	100
5-METHOXYTRYPTAMINE	8.0		306	100
MELATONIN	12.3		360	100

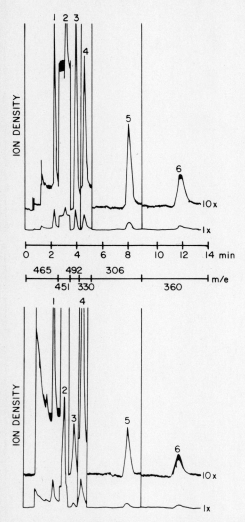

FIG. 4. *Upper tracing:* a record obtained from authentic compounds prepared as described in the text. Time scale is in minutes; the tissues (m/e mass-to-charge) refer to the setting of the magnetic field during that particular time. The peak numbers refer to the PFP derivatives recorded 1. α-methylserotonin, 12 pmole; 2. serotonin, 3.5 pmole; 3. N-acetylserotonin, 3.6 pmole; 4. N-acetyltryptamine, 10 pmole; 5. 5-methoxytryptamine, 3.7 pmole; 6. melatonin, 3.2 pmole. *Lower tracing:* a record obtained from the reacted pineal extract, to which the two internal standards were added.

indole reacted. In this way the linear response obtained (Fig. 5) is used to quantitate endogenous indole levels.

Endogenous indolealkylamines are extracted by homogenizing the tissue in 0.1 M $ZnSO_4$. Following neutralization with 0.1 M $Ba(OH)_2$ and centrifugation, an aliquot of the supernatant is combined with internal standards and dried under a stream of nitrogen. The dried extract is then reacted with 100 µl PFPA in 20 µl ethyl acetate. After 3 hr at 60°C the excess PFPA is driven off under a stream of nitrogen and the residue reconstituted in 10

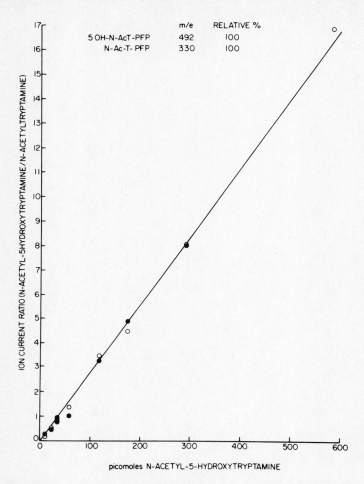

FIG. 5. Calibration curve for N-acetylserotonin (5 hydroxy-N-acetyltryptamine). The internal standard N-acetyltryptamine was kept constant and the N-acetylserotonin concentration varied. Using peak heights as the measurement, the indole/internal standard ratio (ordinate) is plotted against the known concentrations (abscissa) of the indole. These analyses were performed on 2 separate days, open versus closed circles.

μl ethyl acetate. Usually, 2 μl of this solution is injected into the GC port. For the pineal gland, 50 μl of $ZnSO_4$ and 50 μl of $Ba(OH)_2$ are used to extract one pineal gland, and 60 to 75 μl of the supernatant is analyzed by mass fragmentography (Fig. 4).

II. RESULTS AND DISCUSSION

The quantities of the four indoles present in the pineal gland are given in Table 2. The values for S and NAS are in good agreement with the values reported for these substances using either ninhydrin (Neff, Barrett, and Costa, 1969) or O-phthaldehyde (OPT) (Miller and Maickel, 1970).

TABLE 2. *Mass fragmentographic analysis of indole content of rat pineal gland*

Indole	Concentration nmole/g (mean + SE)	Multiple-ion detection (fragment ratio)	
		Tissue extract	Authentic compound
Serotonin	307 ± 25	2.3	2.3
5-Methoxytryptamine	22 ± 2.0	1.0	1.0
N-Acetylserotonin	.60 ± .10	14.3	14.3
Melatonin	16.8 ± .43	2.1	2.1

Animals were housed in an environment of 12 hr light and 12 hr dark, and were sacrificed 1 week after acclimatization at 2 hr of light. Four individual pineals were analyzed as described in the text.

The quantity of M and 5MT as measured by GC-MS is in disagreement with the amount reported by Miller and Maickel (1970) using the OPT method. The specificity of the OPT method is based on selective extraction since the fluorescent characteristics of the reaction products of OPT with the various indoles are identical (Maickel and Miller, 1966). It is, therefore, obvious that any cross–contamination of any one indole into the "specific" extract of a second indole will give erroneous values. Specificity of the GC-MS method is based on GC retention time and multiple-ion detection (MID). MID takes advantage of the fact that the fragmentation pattern of any one compound is completely reproducible in terms of the relative intensity of each of the fragments to each other. Since we are working with low concentrations it is impossible to take a complete mass spectrum of the compound. It is possible, however, to simultaneously monitor two or three characteristic fragments and compare the relative intensities of these fragments to each other. These intensity ratios are then compared to the ratios obtained from authentic compounds and only when these ratio values are in agreement is identification considered absolute. The fragments monitored for MID of the four indoles are shown in Fig. 6, together with their relative abundance and fragment ratios. By subjecting the pineal extract to MID analysis, confirmation of the compounds was obtained (Table 2).

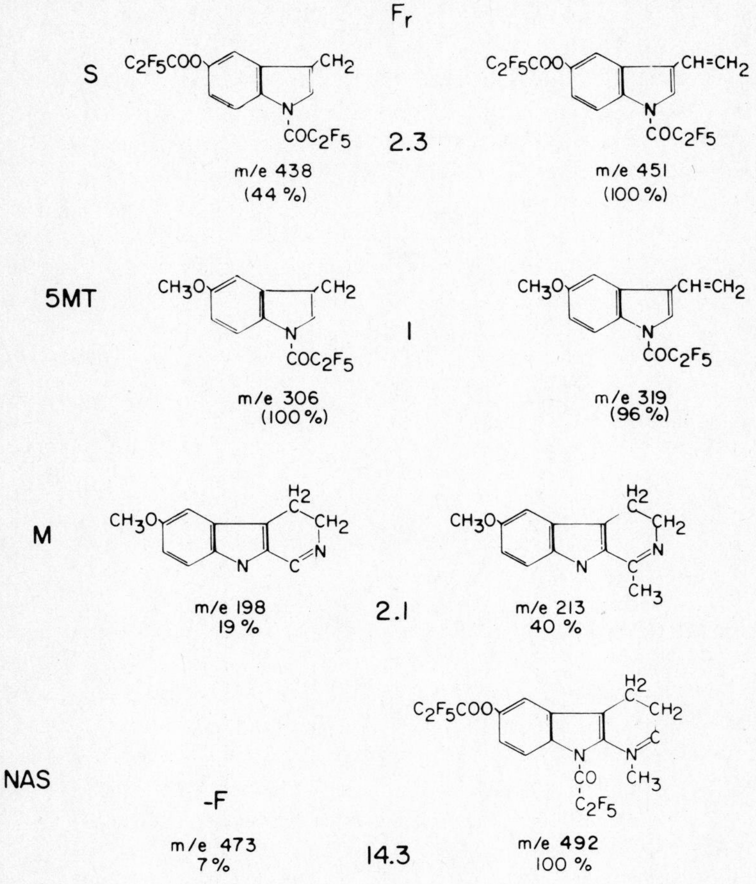

FIG. 6. Positive ions monitored for multiple-ion detection analysis. F_r, fragment ratio; m/e, mass-to-charge; relative percent abundance as compared to base peak.

For several years, serotonin has been considered to be a putative neurotransmitter in the brain. In the last 2 years, reports have appeared suggesting that, in addition to serotonin, other indolealkylamines may be present in the CNS neurons. Björklund, Falck, and Stenevi (1971a, b), using microspectrofluorometric techniques, have found evidence for the existence of an indole of unknown structure in neurons of the rat mesencephalon. The report by Costa and Revuelta (1972) that p-chloroamphetamine is less active in depleting the S content of the hypothalamus and brainstem than that of the telencephalon and diencephalon adds to the speculation that an indole other than S may exist in the hypothalamus and brainstem. In addition, studies by Spano and Costa (1972) showed the presence of an indole other than S in rat hypothalamus.

In an attempt to isolate and identify this other indole, we analyzed hypothalamic extracts by MID for 5MT, NAS, and M and, in addition, tested these extracts for the presence of S (Green, Koslow, and Costa, 1973). The results of the MID analysis of rat hypothalamus confirmed the existence of S and conclusively showed the presence of 5MT (Table 3). NAS was not detected. The MID analysis for M was inconclusive. The molecular ion (m/e 360) which is also the base peak (100%) was recorded at the retention time for M. The height of this peak in hypothalamic homogenates was approximately 15 mm at maximal amplification. Other fragments that were measured for MID were m/e 213 (40%) and m/e 198 (20%). Since it was difficult to discern these small responses (a few mm) from the background noise, we can only tentatively report the presence of M. Absolute identification requires the measurement of at least two fragments and the verification of the known fragment ratio.

TABLE 3. *Mass fragmentographic analysis of indole content of normal and pinealectomized rat hypothalamus*

Indole	Normal	Pinealectomized	Multiple-ion detection (fragment ratio)
Serotonin	3.96 ± 0.23 (8)	3.63 ± 0.51 (3)	2.3
5-Methoxytryptamine	0.62 ± 0.02 (8)	0.46 ± 0.07 (3)	1.0
N-Acetylserotonin	Not detected	Not detected	—
Melatonin	1.5 ± 0.16 (8)	1.4 ± 0.20 (3)	*

Animals were housed in a light/dark cycle of 12/12, and sacrificed at 2 hr of light. Pinealectomy was performed 21 to 28 days prior to analysis. Concentrations are given as nmole/g (mean ± SE). Number of determinations is shown in parentheses.* Melatonin concentrations are based on detection of the molecular ion (m/e 360) at the retention of authentic melatonin.

The concentrations of hypothalamic S, 5MT, and M as analyzed by GC-MS are given in Table 3. Since the hypothalamus is capable of taking up M from both the CSF and blood (Anton-Tay and Wurtman, 1969) and M is synthesized in the pineal gland, it was necessary to exclude the possibility that the hypothalamic M and 5MT are originating from the pineal gland. The hypothalamus of pinealectomized animals were analyzed 21 to 28 days after surgery (Table 3). Since there are no significant differences in the concentrations of S, 5MT, and M in hypothalami of normal and pinealectomized rats, it is clear that these compounds are not originating in the pineal.

Whether they are synthesized in the hypothalamus remains to be proven because we have also found 5MT and M in rat blood.

Björklund et al. (1971*b*) showed that the indole compound they identified has fluorescence spectral characteristics similar to 5MT, 6-hydroxytrypta-mine, or N-methylserotonin. We have examined rat hypothalamus for N-methylserotonin and, thus far, our results have been negative. If the 5MT we measured in the hypothalamus is the same compound that Björk-lund found in nerve cell bodies and terminals, then it is interesting to speculate that 5MT may be a neurotransmitter in the CNS. Indeed, using microiontophoretic techniques, Bradley (1972) has reported that 5MT mimics the effect of S on brainstem neutrons. These small bits of evidence are inconclusive to designate a neurotransmitter role to 5MT. In fact, it may merely be present as a metabolite of S or possibly a precursor in the synthetic pathway of M.

III. SUMMARY

In conclusion, the GC-MS method for S, 5MT, NAS, and M has a sensi-tivity in the order of $10^{-13} - 10^{-14}$ mole. Specificity is routinely based on measurement of the ion density of the characteristic fragment of the com-pound at the substance's retention time. Specificity is based on MID, by comparing the fragment ratios obtained from the extracted compound to those obtained from pure standards. The method of isolation is rapid, and, following derivatization, one analysis may be run every 12 to 15 min, yield-ing the determination of four compounds per GC-MS run.

We have used this method to analyze pineal indoles and investigate the identity of CNS indoles. We have shown the absolute presence of S and 5MT in rat hypothalamus and blood, and tentatively identified M in these two tissues. Our measurements do not rule out the existence of other indoles being present because, based on reports in the literature (Narasim-hachari and Himwich, 1972; Saavedra and Axelrod, 1972), we should still consider the possible existence of N,N-dimethylserotonin and other yet to be identified indoles.

REFERENCES

Anton-Tay, F., and Wurtman, R. J. (1969): Regional uptake of ³H-melatonin from blood or cerebrospinal fluid by rat brain. *Nature*, 221:474–475.

Björklund, A., Falck, B., and Stenevi, U. (1971*a*): Classification of monoamine neurones in the rat mesencephalon: Distribution of a new monoamine system. *Brain Research*, 32:269–285.

Björklund, A., Falck, B., and Stenevi, U. (1971*b*): Microspectrofluorimetric characterization of monoamines in the central nervous system: Evidence for a new neuronal monoamine-like compound. *Progress in Brain Research*, 34:63–73.

Bradley, P. B. (1972): Mechanism of action of psychotomimetic drugs. In: *Fifth International Congress on Pharmacology*, p. 167. San Francisco, California.

Cattabeni, F., Koslow, S. H., and Costa, E. (1972): Gas chromatographic-mass spectrometric assay of four indole alkylamines of rat pineal. *Science*, 178:166–168.

Costa, E., and Revuelta, A. (1972): (-) *p*-Chloroamphetamine and serotonin turnover in rat brain. *Neuropharmacology*, 11:291–295.

Falck, B., Hillarp, N. Å., Thieme, G., and Torp, A. (1962): Fluorescence of catecholamines and related compounds condensed with formaldehyde. *Journal of Histochemistry and Cytochemistry*, 10:348–354.

Green, A. R., Koslow, S. H., and Costa, E. (1973): Identification and quantitation of a new indolealkylamine in rat hypothalamus. *Brain Research*, 51:371–374.

Koslow, S. H., Cattabeni, F., and Costa, E. (1972): Norepinephrine and dopamine: Assay by mass fragmentography in the picomole range. *Science*, 176:177–189.

Koslow, S. H., Cattabeni, F. and Costa, E. (1973): Quantitative mass fragmentography of some indolealkylamines of the rat pineal glands. In: Pineal Gland: Proceedings of a Workshop held by the National Institutes of Health.

Maickel, R. P., and Miller, F. P. (1966): Fluorescent products formed by reaction of indole derivatives and O-phthalaldehyde. *Analytical Chemistry*, 38:1937–1938.

Miller, F. P., and Maickel, R. P. (1970): Fluorometric determination of indole derivatives. *Life Sciences*, 9:747–752.

Narasimhachari, N., and Himwich, H. E. (1972): The determination of bufotenin in urine of schizophrenic patients and normal controls. *Journal of Psychiatric Research*, 9:113–121.

Neff, N. H., Barrett, R. E., and Costa, E. (1969): Kinetic and fluorescent histochemical analysis of the serotonin compartments in rat pineal gland. *European Journal of Pharmacology*, 5:348–356.

Saavedra, J. M., and Axelrod, J. (1972): Psychotomimetic N-methylated tryptamines: Formation in brain *in vivo* and *in vitro*. *Science*, 175:1365–1366.

Spano, P. F., and Costa, E. (1972): Ring substituted indolealkylamines present in rat hypothalamus. *Fifth International Congress on Pharmacology*, p. 1315. San Francisco, California.

Advances in Biochemical Psychopharmacology, Vol. 7
Raven Press, New York © 1973

Use of Mass Fragmentography to Detect Subcellular Localization of Drug Metabolites Acting as False Transmitters

F. Cattabeni, G. Racagni, and A. Groppetti*

Institute of Pharmacology and Pharmacognosy, School of Pharmacy, University of Milano, 20129-Milano, Italy

I. INTRODUCTION

The false neurotransmitter role of catecholamine-like compounds has been established in peripheral nervous tissue of animals treated with the appropriate precursors (review by Kopin, 1968). p-Hydroxynorephedrine (p-OHNE), a metabolite of D-amphetamine (D-A), has also been included among these false transmitters. In fact p-OHNE is formed *in vivo* in nervous tissue (Goldstein and Anagnoste, 1965), and replaces norepinephrine (NE) in the storage granule (Brodie, Cho, and Gessa, 1970; Costa and Groppetti, 1970). It can then be released under conditions known to release NE, such as electrical sympathetic stimulation (Fisher, Horst, and Kopin, 1965; Thoenen, Hürliman, Gey, and Haefely, 1966) or treatment with reserpine or successive doses of D-A (Brodie et al., 1970; Costa and Groppetti, 1970). Further evidence for the false neurotransmitter role of p-OHNE in peripheral nervous tissue was obtained by submitting animals predosed with tritiated A to either cold exposure, a situation known to increase the turn-over rate of cardiac NE (Oliverio and Stjärne, 1965; Gordon, Spector, Sjoerdsma, and Udenfriend, 1966), or treatment with desmethylimipramine, an inhibitor of amine uptake by sympathetic axons (Groppetti and Costa, 1968a). In both cases, cardiac p-OHNE disappeared faster than in control animals, suggesting therefore that p-OHNE is released and taken up by a mechanism similar to that involved for NE. However, desmethylimipramine treatment and cold exposure failed to increase the disappearance rate of brain p-OHNE, probably because the blood-brain barrier prevents the efflux

---

* Institute of Pharmacology, School of Medicine, 2nd Chair, University of Milano, 20129 – Milano, Italy

of such a polar compound which is also resistant to enzymatic degradation (Groppetti and Costa, 1968*a*).

Radioautographic techniques have been unsuccessfully applied to this problem, because only 10% of the radioactivity present was associated with *p*-OHNE, the remaining tritium being associated with tissue water and other brain components (Groppetti and Costa, 1968*b*).

The clarification of the central role of *p*-OHNE is important, since it may well be that brain accumulation of this metabolite is responsible for the psychotic effects in the chronic use of A. This chapter describes our preliminary findings with mass fragmentography in an attempt to shed new light on this question. (The principles of mass fragmentography are presented in this volume in the chapter by Holmstedt and Palmér).

In conjunction with this investigation, we have continued our studies dealing with the mechanism of action of fenfluramine (N-ethyl-α-methyl-3-trifluoromethyl-phenylethylamine). This anorectic, nonstimulating drug has been shown to selectively reduce rat brain 5-hydroxytryptamine (5-HT) (Duhault and Verdavainne, 1967).

We have previously observed (Morgan, Cattabeni, and Costa, 1972) the rapid *in vivo* dealkylation of fenfluramine to norfenfluramine. Our working hypothesis is that norfenfluramine, although structurally unrelated to 5-HT, can act as a false neurotransmitter in the serotonergic system. No other false transmitters have so far been found for this system, except for some indications suggesting that after a massive dose of L-DOPA, the dopamine formed displaces 5-HT from its storage sites and possibly is released by electrical stimulation (Butcher, Engel, and Fuxe, 1970; Ng, Colburn, and Kopin, 1971).

II. *p*-HYDROXYNOREPHEDRINE

A. Identification and Assay of *p*-OHNE by Mass Fragmentography

Identification and assay of D-A metabolites have been based upon radio-isotopic techniques in combination with thin-layer and paper chromatography. These methods are not only relatively nonspecific, but also have the disadvantages that with tritiated D-A, the tritium ions can exchange with tissue water and this leads to incorrect results if the quantitative assay is based on radioactivity measurements. Hence, we have developed a quantitative mass fragmentographic assay for *p*-OHNE, which permits measurement of picomole amounts in tissue samples.

Briefly, the tissue is homogenized in 10 volumes of acidic butanol (0.85 ml of concentrated hydrochloric acid in 1.000 ml of n-butanol) and centrifuged at 10,000 rpm for 15 min. An aliquot is transferred, and 0.2 to 0.5 ml of 0.1 M formic acid and 2 volumes of n-heptane are added (Shore and Olin, 1958), in order to increase the solubility of the amines in the aqueous phase. The organic phase is discarded and the aqueous phase is washed twice with 3 volumes of ethyl ether. An aliquot (50 to 100 μl) of the aqueous phase is then transferred in a 0.3-ml vial containing a known amount (50 pmole) of the internal standard, α-methyldopamine (α-MDM). The sample is dried under a stream of nitrogen and heated to 60°C for at least 30 min with 100 μl of pentafluoropropionic anhydride in the presence of 20 μl of ethyl acetate. The excess reagent is evaporated under nitrogen prior to analysis and the sample is reconstituted with 10 μl of ethyl acetate and 1 to 2 μl is injected into the gas chromatograph-mass spectrometer LKB 9000 (conditions in Table 1).

TABLE 1. *Gas-chromatographic—mass spectrometric characteristics for amphetamine and metabolites*

$$R^{11} \quad CH_3$$
$$R^1 \quad | \qquad |$$
$$CH—CH—NH_2$$
$$\beta \qquad \alpha$$
$$R$$

Compounds*	RRT (αMDM)	Substituents			m/e of α-β carbons cleavage fragment**(%)
		R	R^1	R^{11}	
Norephedrine-PFP	0.36	H	H	OH	190 (100)
Amphetamine-PFP	0.43	H	H	H	190 (100)
pOH-Norephedrine-PFP	0.48	OH	H	OH	190 (100)
pOH-Amphetamine-PFP	0.77	OH	H	H	190 (100)
α-Methyldopamine-PFP (αMDM, Internal standard)	1	OH	OH	H	190 (100)

Conditions: GC-MS LKB 9000; Column: 4 mt., OV 17 3%, OVEN TEMP. 150°C, HELIUM FLOW 40 ml/min. ION SOURCE TEMP. 270°C, ELECTRON ENERGY 40eV TRAP CURRENT 60 μA.
*PENTAFLUOROPROPIONYL DERIVATIVES (PFP)
**THE POSITIVE CHARGE IS RETAINED ON THE NITROGEN-CONTAINING FRAGMENT.

Due to the paucity of characteristic fragments in the spectra of the pentafluoropropionyl derivatives (PFP) of A and its metabolites (Fig. 1), we have been forced to use single-ion detection rather than multiple-ion detection for the identification of p-OHNE in brain of D-A-treated rats. The appearance of a peak with the gas chromatograph-mass spectrometer focused on the most abundant fragment (mass number 190, originated by α-β carbon

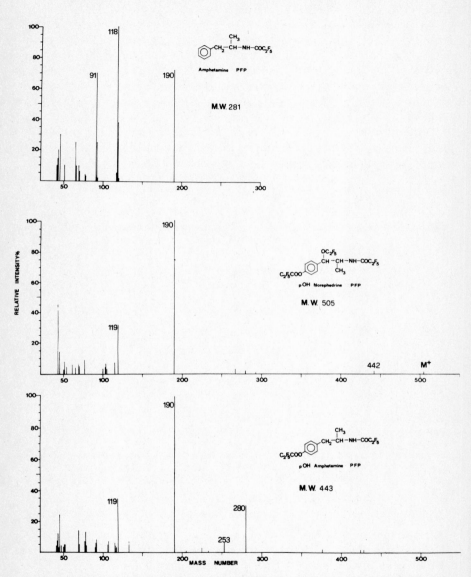

FIG. 1. Mass spectra of pentafluoropropionyl derivatives (PFP) of amphetamine, *p*-hydroxyamphetamine, and *p*-hydroxynorephedrine. For mass spectrometric conditions see Table 1.

bond cleavage with the charge retained on the nitrogen-containing part), together with the retention time relative (RRT) to α-MDM, served as criteria for the identification of *p*-OHNE (Table 1).

A mass fragmentogram obtained from a brain extract of a D-A-treated rat is represented in Fig. 2C, together with that of a brain extract obtained from a control animal (Fig. 2A) and authentic compounds (Fig. 2B). The gas chromatograph-mass spectrometer was focused upon mass number 190 and the peak with RRT of 0.48 corresponded to that obtained for authentic *p*-OHNE.

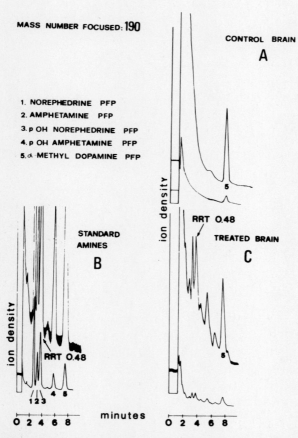

FIG. 2. Mass fragmentogram obtained from analysis of A: brain extract of a control rat; B: authentic amphetamine and metabolites, pentafluoropropionyl derivatives; and C: brain extract of an animal pretreated with 7 mg/kg i.p. of D-amphetamine. The peak with RRT 0.48 corresponds to *p*-hydroxynorephedrine. For conditions see Table 1

The *p*-OHNE concentrations measured with mass-fragmentography in brain and heart of rats given 7 mg/kg i.p. of D-A, 1 and 3 hr after treatment (Table 2) are in good agreement with those obtained with radioisotopic techniques (Brodie et al., 1970; Costa and Groppetti, 1970). The slightly lower values obtained with the latter method could be explained, as previously mentioned, with the exchange of the A bound tritium with the tissue water.

TABLE 2. p-*Hydroxynorephedrine concen-trations in brain and heart of rats*

Hr of treatment	p-Hydroxynorephedrine ng/g ± SE	
	Brain	Heart
1	62 (2)	325 ± 33 (3)
3	257 ± 25 (3)	507 ± 107 (3)

D-Amphetamine: 7 mg/kg i.p.
Number of animals in parentheses.

B. Formation and Localization of *p*-OHNE in Noradrenergic Storage Sites

Since it has been demonstrated that *p*-hydroxyamphetamine, the first metabolic product of A in rats (Axelrod, 1954), is a good substrate for dopamine-β-hydroxylase (DBH) in peripheral nervous tissue (Goldstein and Anagnoste, 1965), it has been postulated that *p*-OHNE is formed in the brain within the noradrenergic granules where DBH is located. Accordingly, by destroying the noradrenergic nerve terminals with 6-hydroxy-dopamine (6-OHDM) (Thoenen and Tranzer, 1968), it should be possible to prevent the formation of any β-hydroxylated compound.

To test this possibility, 200 μg/rat of 6-OHDM was administered intra-ventricularly and after 1 week the rats were treated with 7 mg/kg i.p. of D-A and sacrificed 5 hr later. *p*-OHNE concentration in 6-OHDM-treated animals was decreased to 36% when compared with control animals (Fig. 3). It is interesting to note that the percent NE decrease in these conditions has approximately the same value (35%); suggesting that the amount of *p*-OHNE formed in brain is directly related to the activity of the noradren-ergic system. The concentration of A in brain is not altered after 6-OHDM treatment, thus excluding that a lower amount of the substrate is responsible for the lower concentration of *p*-OHNE.

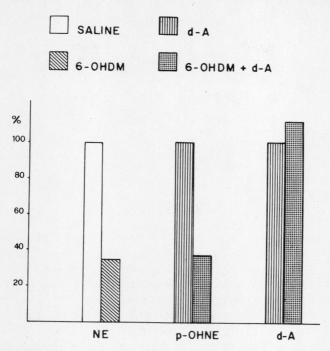

FIG. 3. Impaired formation of *p*-hydroxynorephedrine (*p*-OHNE) in brain of rats treated with 6-hydroxydopamine (6-OHDM, 200 μg/rat, intraventricularly) and 1 week later with D-amphetamine (7 mg/kg i.p.). Brain *p*-OHNE concentrations are compared with those of norepinephrine (NE) and amphetamine (A).

In order to further substantiate the relationship between the noradrenergic system and *p*-OHNE formation and storage, we have determined the concentration of the D-A metabolite in brain areas characterized by a low or a high population of noradrenergic nerve terminals, such as the caudate nucleus and the hypothalamus, respectively. Five hr after i.p. administration of 7 mg/kg of D-A the hypothalamic concentration of *p*-OHNE was 10 times higher than that found in the caudate nucleus (420 ng/g versus 40ng/g, Table 3). These results lend further support to the conclusion that brain *p*-OHNE is indeed associated with the noradrenergic system, and we are now investigating whether the subcellular localization of *p*-OHNE is affected by noradrenergic stimulation. Such an experiment should provide additional evidence for the false neurotransmitter role of *p*-OHNE and shed more light on the action of A at the cellular molecular level.

TABLE 3. p-*Hydroxynorephedrine concentra-*
tions in different brain areas of rat

Tissue	p-Hydroxynorephedrine ng/g ± SE
Caudate nucleus	40.0 ± 3.8 (3)
Hypothalamus	420.7 ± 72.1 (3)
Rest of the brain	179.8 ± 30.0 (3)

D-Amphetamine: 7 mg/kg i.p., 5 hr treatment.
Number of animals in parentheses.

III. NORFENFLURAMINE

The mechanism of the long-lasting depletion of 5-HT elicited by fenfluramine seems to be correlated with the presence of the N-dealkylated metabolite norfenfluramine (Morgan et al., 1972). After norfenfluramine treatment, the *in vivo* conversion of labeled tryptophan into 5-HT is greater
than normal (Costa and Revuelta, 1972) and neither tryptophan hydroxylase
nor aromatic amino acid decarboxylase is inhibited (Morgan, Löfstrandh,
and Costa, 1972), suggesting that norfenfluramine acts on serotonergic
storage sites. It was, therefore, of interest to investigate the subcellular
localization of norfenfluramine, in order to explain the long-lasting depletion
of 5-HT elicited by this drug.

A. Subcellular localization of norfenfluramine

A quantitative mass fragmentographic assay has been developed which
is capable of measuring minute quantities of norfenfluramine in subcellular
fractions of rat brain (Cattabeni, Racagni, and Costa, 1973). Single-ion detection was employed as before and fenfluramine served as the internal
standard. Rats were given 90 μmole/kg i.p. of DL-norfenfluramine (Les
Laboratoires Servier, Paris) 5 hr before sacrifice. The brain was removed
and the telencephalon dissected, since this is the region most affected by
this drug with regard to 5-HT depletion (Costa and Revuelta, 1972). The
homogenates were fractionated using high-speed centrifugation and a discontinuous sucrose gradient according to Gray and Whittaker (1962). Each
fraction was made alkaline and extracted with ethyl ether. The ether was
evaporated and the residue treated with pentafluoropropionic anhydride,
as described previously for p-OHNE. Once again, the gas chromatograph-
mass spectrometer was focused upon mass number 190, the most abundant

fragment in the norfenfluramine spectrum (Cattabeni et al., 1973). The values of norfenfluramine concentrations in the different fractions are shown in Table 4. The figures are given as absolute content of norfenfluramine in each fraction, as percent of norfenfluramine in the whole brain, and as concentration per milligram of proteins.

TABLE 4. *Subcellular distribution of norfenfluramine in telencephalon of rat*

Fractions	Norfenfluramine (nmole)	Homogenate (% of total)	Protein (nmole/mg)
Total homogenate	83.81	—	1.58
126,000 × g Supernatant	15.35	18.37	1.97
126,000 × g Pellet (microsomal fraction)	1.21	1.44	0.98
12,000 × g Pellet	25.12	29.97	1.41
0.32 M Sucrose band	0.53	0.63	2.41
Synaptosomal fraction	14.48	17.27	2.44
Mitochondria (pellet)	1.31	1.56	0.29

Each value represents the mean of three experiments.
90 μmole/kg of norfenfluramine were administered 5 hrs before sacrifice i.p.

The crude mitochondrial fraction (12,000 × g pellet) contained approximately 25% of the drug found in total homogenate. This fraction was submitted to high-speed centrifugation on discontinuous sucrose gradient (0.32–0.6–0.8–1–1.2 M) and three fractions were collected. The first fraction (0.32 M sucrose band), composed of synaptic vesicles and some synaptosomes, contained very little norfenfluramine (0.53%). If the content of protein was considered, however, this fraction shows an enrichment of concentration in protein, i.e., 2.41 nmole/mg protein versus 1.58 nmole/mg of the total homogenate. A similar concentration (2.44) was obtained in the pooled synaptosomal fractions (0.6–0.8–1 M) where 17.3% of the drug is present. The amount of norfenfluramine located in the mitochondrial pellet was very low considering either the amount of drug (1.56%) or its concentration in protein (0.29 nmole/mg). The supernatant of the 12,000 × g centrifugation was spun at 126,000 × g for 60 min. Together, the supernatant and microsomal fraction contained 16.5% of the drug. The concentration in protein of these fractions was lower (microsomal fraction) or equal (supernatant) to that of total homogenate.

The above results suggest that norfenfluramine is localized in synaptosomes and we are now investigating whether this localization is specific for serotoninergic storage sites. The long-lasting depletion of telencephalic 5-HT in rats seems to be due to the accumulation of norfenfluramine within the granules, and it is tempting to speculate that this drug could be released under stimulation and act as a false transmitter.

IV. SUMMARY

A sensitive and specific mass fragmentographic method for the assay of picomole amounts of *p*-hydroxynorephedrine in tissue samples has been developed. Rats receiving D-amphetamine accumulate *p*-hydroxynorephedrine in brain and other tissues, in agreement with previous reports. The concentration of this metabolite seems to be proportional to tissue levels of norepinephrine. In contrast to amphetamine, *p*-hydroxynorephedrine is not homogeneously distributed in the different brain areas: *p*-hydroxynorephedrine is mainly concentrated in the hypothalamus, whereas only trace amounts of this metabolite have been found in the caudate nucleus. *p*-Hydroxynorephedrine does not accumulate in brain of rats after chemical sympathectomy with an intraventricular dose of 6-hydroxydopamine. These results indicate that in rat brain *p*-hydroxynorephedrine is selectively retained and formed by some specialized mechanisms related to the function of noradrenergic neurones.

Subcellular localization studies show that the nonstimulating, anorectic compound norfenfluramine is mainly localized in the synaptosomal fraction. This localization suggests that the long-lasting depletion of brain 5-hydroxytryptamine elicited by this drug is due to its accumulation in the serotoninergic storage sites. We speculate also that norfenfluramine can be released from the storage granules and act as a false transmitter.

REFERENCES

Axelrod, J. (1954): Studies on sympathomimetic amines. II. The biotransformation and physiological disposition of D-amphetamine, D-*p*-hydroxyamphetamine and D-methamphetamine. *Journal of Pharmacology and Experimental Therapeutics,* 110:315.

Brodie, B. B., Cho, A. K., and Gessa, G. L. (1970): Possible role of *p*-hydroxynorephedrine in the depletion of norepinephrine induced by D-amphetamine and in tolerance to this drug. In: *International Symposium on Amphetamines and Related Compounds,* edited by E. Costa and S. Garattini, pp. 217–230. Raven Press, New York.

Butcher, L. L., Engel, J., and Fuxe, C. (1970): L-DOPA induced changes in central monoamine neurons after peripheral decarboxylase inhibition. *Journal of Pharmacy and Pharmacology,* 22:313.

Cattabeni, F., Racagni, G., and Costa, E. (1973): Methamphetamine, fenfluramine and their metabolites: Identification and subcellular localization in rat brain homogenates. In: *Advances in Neuropsychopharmacology–1972,* AVICENUM, Czechoslovak Medical Press, Praha, *in press.*

Costa, E., and Groppetti, A. (1970): Biosynthesis and storage of catecholamines in tissues of rats injected with various doses of *d*-amphetamine. In: *International Symposium on Amphetamines and Related Compounds,* edited by E. Costa and S. Garattini, pp. 231–255. Raven Press, New York.

Costa, E., Groppetti, A., and Revuelta, A. (1971): Action of fenfluramine on monoamine stores in rat tissues. *British Journal of Pharmacology*, 41:57.

Costa, E., and Revuelta, A. (1972): Norfenfluramine and serotonin turnover rate in the rat brain. *Biochemical Pharmacology*, 21:2385.

Duhault, J., and Verdavainne, C. (1967): Modification du taux de sèrotonine cérébral chez le rat par les trifluorométhyl-phényl-2-éthyl aminopropane (fenfluramine 768-S). *Archives Internationaux de Pharmacodynamie et de Thérapie*, 170:276.

Fisher, J. E., Horst, W. D., and Kopin, I. J. (1965): β-Hydroxylated sympathomimetic amines as false neurotransmitters. *British Journal of Pharmacology*, 24:477.

Goldstein, M., and Anagnoste, B. (1965): The conversion *in vivo* of d-amphetamine to (+)-p-hydroxynorephedrine. *Biochimica Biophysica Acta*, 107:166.

Gordon, R., Spector, S., Sjoerdsma, A., and Udenfriend, S. (1966): Increased synthesis of norepinephrine and epinephrine in the intact rat during exercise and cold exposure. *Journal of Pharmacology and Experimental Therapeutics*, 153:440.

Gray, E. G., and Whittaker, V. P. (1962): The isolation of nerve endings from brain: An electron microscopic study of cell fragments derived by homogenization and centrifugation. *Journal of Anatomy*, 96:79.

Groppetti, A., and Costa, E. (1968a): Effects of cold exposure and DMI on the p-hydroxynorephedrine-^{3}H concentrations in tissues of rats injected with d-amphetamine-^{3}H. *Atti Accademia Medica Lombarda*, 23:1105.

Groppetti, A., and Costa, E. (1968b): Release of p-hydroxynorephedrine-^{3}H by d-amphetamine-^{3}H. *Atti Accademia Medica Lombarda*, 23:1113.

Kopin, I. J. (1968): False adrenergic transmitters. *Annual Review of Pharmacology*, 8:377.

Morgan, C. D., Cattabeni, F., and Costa, E. (1972): Methamphetamine, fenfluramine and their N-dealkylated metabolites: effect on monoamine concentrations in rat tissues. *Journal of Pharmacology and Experimental Therapeutics*, 180:177.

Morgan, C. D., Löfstrandh, S., and Costa, E. (1972): Amphetamine analogues and brain amines. *Life Sciences*, 11:83.

Ng, K. Y., Colburn, R. W., and Kopin, I. J. (1971): Effects of L-DOPA on efflux of cerebral monoamines from synaptosomes. *Nature*, 230:331.

Oliverio, A., and Stjärne, L. (1965): Acceleration of noradrenaline turnover in the mouse heart by cold exposure. *Life Sciences*, 4:2339.

Shore, P. A., and Olin, J. S. (1958): Identification and chemical assay of norepinephrine in brain and other tissues. *Journal of Pharmacology and Experimental Therapeutics*, 122:295.

Thoenen, H., Hürliman, K. F., Gey, K. F., and Haefely, W. (1966): Liberation of p-hydroxynorephedrine from cat spleen by sympathetic nerve stimulation after pretreatment with amphetamine. *Life Sciences*, 5:1715.

Thoenen, H., and Tranzer, J. P. (1968): Chemical sympathectomy by selective destruction of adrenergic nerve endings with 6-hydroxydopamine. *Archives für Pharmakologie und Experimentelle Pathologie*, 261:271.

Advances in Biochemical Psychopharmacology, Vol. 7
Raven Press, New York © 1973

The Use of Stable Oxygen Isotopes for Labeling of Homovanillic Acid in Rat Brain *In Vivo*

Göran Sedvall, Avraham Mayevsky, Claes-Göran Fri, Birgitta Sjöquist, and David Samuel

Department of Pharmacology, Karolinska Institute, Stockholm, Sweden, and the Isotope Department, The Weizmann Institute of Science, Rehovot, Israel

I. INTRODUCTION

Brain dopamine is involved in extrapyramidal motor control and the lack of this transmitter is related to the development of Parkinson's disease (Hornykiewicz, 1966). Dopamine is also of importance in behavioral manifestations following the administration of psychoactive drugs such as amphetamine (Costa and Groppetti, 1970) and antipsychotics (Nybäck and Sedvall, 1971). Moreover, malfunctions of dopaminergic transmission have been suggested to occur in schizophrenia (Randrup and Munkvad, 1967). A major fraction of the dopamine in brain is confined to a neuronal pathway extending from the substantia nigra in the midbrain into the corpus striatum (Andén, Roos, and Werdinius, 1964). In such neurons, tyrosine and possibly also phenylalanine are hydroxylated in reactions requiring tyrosine 3-hydroxylase (EC 1.10.3.1), molecular oxygen, and reduced pteridine cofactor (Nagatsu, Levitt, and Udenfriend 1964; Shiman, Akino, and Kaufman 1971). Dihydroxyphenylalanine (DOPA) is formed and then rapidly decarboxylated to dopamine, which is the final transmitter of these neurons (Fig. 1). The main metabolite of dopamine in the brain of most species, including man, is homovanillic acid (HVA) (Hornykiewicz, 1966).

To elucidate the role of dopamine in brain function, a number of techniques have been developed for the quantitative determination of its turnover in the animal brain. Levels of HVA in brain of animals and in lumbar cerebrospinal fluid of man have been claimed to reflect dopamine turnover (Papeschi, 1972). However, since mechanisms for HVA transport between various body compartments are poorly understood and not readily controlled, other procedures have been devised. In animals the use of cate-

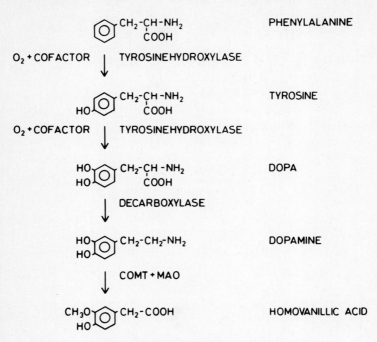

FIG. 1. Pathway of dopamine synthesis and metabolism.

cholamine synthesis inhibitors or precursors labeled with radioactive isotopes seems to be the most valid procedure developed so far for quantitative determination of brain catecholamine turnover (Costa, 1972). For studies in man, however, there is limited use for such methods due to the toxicity of pharmacological tools and radiation from radioactive isotopes.

The recent development of gas chromatographic and mass spectrometric methods for determination of catecholamines and their metabolites (Änggård and Sedvall, 1969; Koslow, Cattabeni, and Costa, 1972, Sjöquist, Dailey, Sedvall, and Änggård, 1973) offers the possibility to use stable isotopes to label brain catecholamines *in vivo*. Thereby the risk of radiation can be avoided. Several stable isotopes are now available for use. Thus precursor amino acids to the catecholamines can be labeled with ^{2}D, ^{13}C, or ^{15}N, and subsequently administered *in vivo*. Since only a small proportion of the precursors goes into the catecholamine pathway, very large amounts have to be administered to make detection possible. Moreover, by mass effects on precursor pools, influences on catecholamine synthesis might be induced.

Since the catecholamines are formed by hydroxylases using molecular oxygen, the possibility of using stable oxygen isotopes for *in vivo* labeling is of considerable interest. Molecular oxygen can easily be administered by inhalation, and by a single exposure it should be possible to pulse label *in vivo* a number of compounds of biological interest.

Homovanillic acid is synthesized in the rat brain following the pathway depicted in Fig. 1. Since steady-state levels of brain dopamine are maintained constant during most physiological and pharmacological conditions, the rates of dopamine synthesis could be determined by following the changes in specific activity of dopamine and HVA, its main metabolite, after pulse labeling of the compounds by inhalation of stable oxygen isotopes. The rationale for primarily labeling HVA is that it is the main dopamine metabolite leaving the brain via the cerebrospinal fluid and blood, where it can be detected before reaching urine, the final pool to which the total body production of dopamine metabolites is delivered. Since a mass spectrometric method for the determination of HVA in rat brain was recently developed in our laboratory (Änggård and Sedvall, 1969, Sjöquist et al., 1973), we have elucidated the possibility of using stable oxygen isotopes to label HVA in rat brain *in vivo*.

II. IN VIVO EXPOSURE OF RATS TO ^{18}O CONTAINING ATMOSPHERES

The proportion of stable oxygen isotopes occurring in nature is shown in Table 1. At the Isotope Separation Plant of the Weizmann Institute of Science in Rehovot, Israel, water is enriched with regard to $H_2{}^{18}O$ to about 95% by distillation. By electrolysis of the water, gas enriched with $^{18}O_2$ to 90 to 95% can be obtained (Samuel, 1962).

To allow the *in vivo* exposure of rats to atmospheres containing $^{18}O_2$, a

TABLE 1. *Relative abundance of stable oxygen isotopes*

Isotope	Percent		
	^{16}O	^{17}O	^{18}O
$H_2{}^{16}O$*	99.759	0.037	0.204
$H_2{}^{18}O$**	$\simeq 4$	$\simeq 0.5$	$\simeq 95$

*Seawater
**^{18}O-enriched water obtained by distillation at the Isotope Separation Plant, the Weizmann Institute of Science, Rehovot, Israel.

closed system was devised by Mayevsky and Samuel for studies on oxidative metabolism. This system, which is depicted in Fig. 2, was used in the
present investigation to study *in vivo* incorporation of ^{18}O in brain HVA
of rats. Rats were introduced into the system (A), which at the beginning of
the experiment contained air at normal pressure. As the system was closed,
oxygen was consumed by the animals and $^{18}O_2$ gas containing approximately
95% ^{18}O was introduced by a pressure-regulated valve (H) adjusted to main-
tain a constant pressure. The atmosphere within the system was recircu-
lated by a peristaltic pump (E) over a CO_2 trap containing KOH (B) and
silica gel (C, D) to remove humidity from the system. The atmosphere,
which at the beginning of the experiment only contained traces of $^{18}O_2$,
contained 90 to 95% $^{18}O_2$ at the end of exposure, which lasted up to 3 hr.
At regular intervals, gas samples were taken from the atmosphere by special
samplers (S) and the percentage ^{18}O was determined by direct mass spectrometry (Mayevsky, Sjöquist, Fri, Samuel, and Sedvall, 1973). Immediately after the end of exposure the animals were removed from the system,

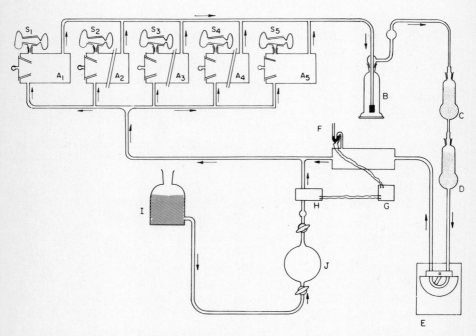

FIG. 2. Chamber for *in vivo* exposure of rats to atmospheres containing $^{18}O_2$. For full explanation see text.

decapitated, and brains were extracted for the determination of ^{18}O incorporation into HVA.

III. MASS SPECTROMETRIC IDENTIFICATION AND DETERMINATION OF HVA IN RAT BRAIN FOLLOWING EXPOSURE TO ^{18}O-CONTAINING ATMOSPHERES

The development of gas chromatographic-mass spectrometric methods for the determination of HVA in tissues and body fluids was a prerequisite for the performance of the present project (Änggård and Sedvall 1969; Sjöquist and Änggård 1973; Sjöquist et al., 1973). The procedure (Fig. 3) involves extraction of brain tissue with ethanol, followed by the addition of deuterated HVA methyl ester as carrier and internal standard, and synthesis of the methyl ester of HVA. After formation of the heptafluorobutyryl derivatives, samples were injected into a combined gas chromatograph-mass spectrometer (LKB 9000). The instrument was equipped with a 1% OV-17 column. Typical mass spectra of extracts from derivatives of authentic HVA and brain obtained from rats exposed for 3 hr to $^{16}O_2$- and $^{18}O_2$-containing atmospheres are depicted in Fig. 4. The derivative of authentic HVA shows major ions at m/e 392 (molecular ion) and 333. The figure demonstrates that, in brain extracts of $^{18}O_2$-exposed animals, mass fragments at the retention time of HVA methyl ester heptafluorobutyryl derivative are obtained at m/e values 2 units higher than those of authentic HVA. This demonstrates the incorporation of one ^{18}O atom in HVA. The data constitute *in vivo* confirmation of the results obtained in enzyme studies by Nagatsu et al. (1964) and Daly, Levitt, Guroff, and Udenfriend (1968), demonstrating the atmospheric origin of the 3-hydroxy group in catecholamines. The mass spectra depicted in Fig. 4 do not demonstrate significant peaks at mass numbers 337 and 396, which indicates that under the experimental conditions, phenylalanine, which is hydroxylated in both the 3 and 4 position by tyrosine hydroxylase (Shiman et al., 1971), did not constitute a major precursor to brain dopamine.

For quantitative determination of the amounts of natural ^{16}O HVA and ^{18}O HVA in rat brain following exposure to $^{18}O_2$-containing atmospheres, the accelerating voltage alternator (AVA) (Sweeley, Elliot, Fories, and Ryhage, 1966) was used. The mass spectrometer was alternately focused on mass numbers 392, 394, and 395, constituting m/e values for the molecular ion of the natural, ^{18}O-labeled, and ^2D-labeled internal standard of the methyl ester heptafluorobutyryl derivatives of HVA respectively (Fig. 5). By comparing peak height ratios with and without the internal standard,

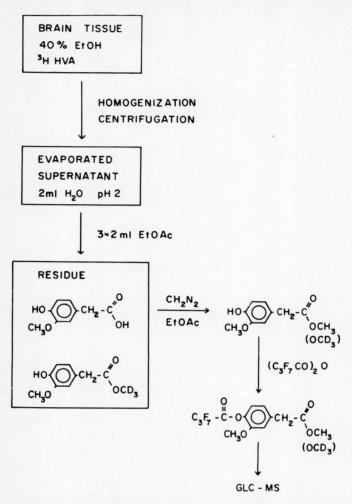

FIG. 3. Flow sheet for extraction and derivatization of HVA from rat brain.

quantification was obtained by means of a standard curve containing known amounts of authentic and ^{2}D-labeled HVA derivatives.

The time course for incorporation of ^{18}O in HVA and the simultaneous disappearance of ^{16}O HVA from rat brain during and after exposure to ^{18}O$_2$-containing atmospheres is presented in Fig. 6. Following the interruption of the exposure, ^{16}O HVA rapidly increases whereas the ^{18}O-labeled form successively disappears.

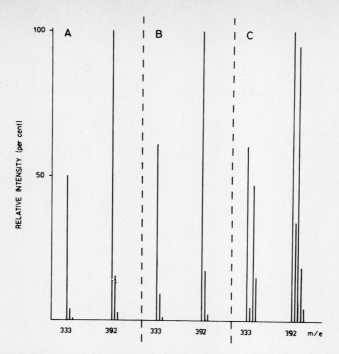

FIG. 4. Partial mass spectrum of methyl ester heptafluorobutyryl derivatives of (A) authentic HVA, (B) HVA from brain of $^{16}O_2$ exposed rats, and (C) HVA from brain of $^{18}O_2$-exposed rats.

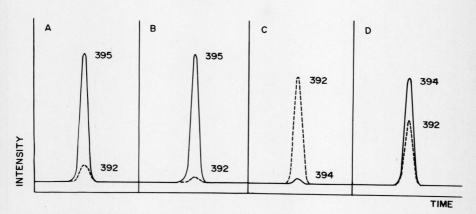

FIG. 5. Mass fragmentogram of methyl ester heptafluorobutyryl derivatives of HVA extracted from brain of (A and C) $^{16}O_2$- and (B and D) $^{18}O_2$-exposed rats. A and B represent extracts with, and C and D without, the addition of the deuterated internal standard. The curve for m/e 395 represents the derivative of the internal standard, 392 represents ^{16}O HVA, and 394 the derivative of ^{18}O HVA.

It is also evident from Fig. 6 that there is no relationship between the accumulation of $H_2{}^{18}O$ in serum water and the content of ^{18}O-labeled HVA. This was also demonstrated by injecting $H_2{}^{18}O$ i.p. into rats, giving contents of $H_2{}^{18}O$ in serum water of the same magnitude as following exposure to $^{18}O_2$ gas. Such an injection did not cause the accumulation of significant amounts of ^{18}O-labeled HVA in the rat brain (Mayevsky et al. 1973). This represents further evidence that the hydroxyl groups in HVA derive from atmospheric oxygen and that the oxygen atoms in the carboxyl group are easily exchangeable by oxygen in water (Smith, Weissbach, and Udenfriend, 1962).

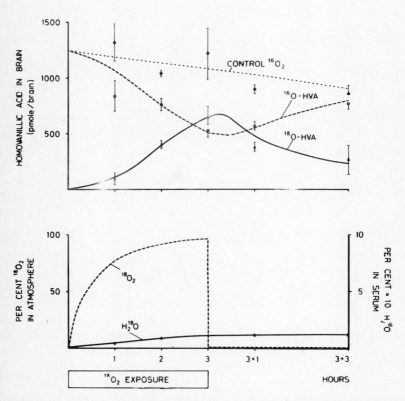

FIG. 6. Time curve for changes in ^{18}O HVA and ^{16}O HVA in brain during and after exposure to $^{18}O_2$.

IV. THE USE OF IN VIVO LABELING WITH ^{18}O TO STUDY CHANGES IN BRAIN DOPAMINE METABOLISM FOLLOWING CHLORPROMAZINE TREATMENT

Previous studies have demonstrated that treatment with chlorpromazine and other types of antipsychotic neuroleptic drugs markedly accelerates brain dopamine synthesis and turnover in the rat brain (Andén, Roos, and Werdinius, 1964; Neff and Costa, 1967; Nybäck and Sedvall, 1968). Therefore, chlorpromazine was used as a tool to study whether the accumulation of ^{18}O-labeled HVA during *in vivo* exposure to ^{18}O$_2$-containing atmospheres reflected dynamic changes in brain dopamine synthesis. Rats were treated with saline or chlorpromazine (10 mg/kg i.p.) and were then immediately introduced into the same exposure chamber. Groups of animals were exposed for 1, 2, or 3 hr to atmospheres containing ^{16}O$_2$ or ^{18}O$_2$. As demonstrated in Fig. 7, chlorpromazine markedly accelerated the rate of accumulation of ^{18}O-labeled HVA in the brain in relationship to saline-treated animals during ^{18}O$_2$ exposure. On the other hand, the fractional rate constant (k) of ^{16}O HVA decline was almost exactly the same in saline and chlorpromazine-treated animals. The k value for ^{16}O HVA disappearance, which seemed to be exponential, was 0.58 ± 0.15 hr^{-1} in saline and 0.58 ± 0.04 hr^{-1} in chlorpromazine-treated rats. This indicates that chlorpromazine does not markedly alter HVA transport from brain, a mechanism that was suggested by Andén et al. (1964) to account for the increase in brain HVA levels following chlorpromazine treatment. The present finding that the rate of ^{18}O HVA but not H$_2$^{18}O (Fig. 7) accumulation in brain increases in chlorpromazine-treated rats exposed to ^{18}O$_2$ in the presence of unaltered k of ^{16}O HVA disappearance, is a direct demonstration that chlorpromazine accelerates brain HVA synthesis. Indirect evidence for this view was previously obtained by the finding that chlorpromazine accelerates synthesis and turnover rate of brain dopamine (Neff and Costa, 1967; Nybäck and Sedvall, 1968). Since a substantial amount of unlabeled brain dopamine was released during ^{18}O$_2$ exposure, the k value for ^{16}O HVA disappearance indicates that a substantial amount of dopamine formed before the beginning of the label with ^{18}O is released during the 3 hr in which the rats are breathing ^{18}O (Wiesel, Fri, and Sedvall, 1973). From the relative differences in rates of ^{18}O incorporation in HVA, it can be calculated that the rate of HVA synthesis in the brain of chlorpromazine-treated animals exceeded by at least threefold that in the brain of saline-treated animals. The efflux rate of ^{16}O HVA is at least twice as fast in chlorpromazine as in saline-treated rats.

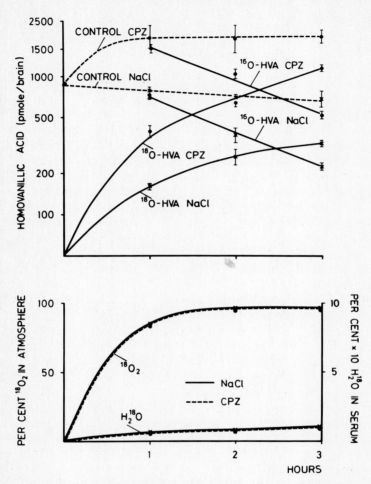

FIG. 7. Effect of chlorpromazine (10 mg/kg i.p.) treatment on changes in rat brain ^{18}O HVA and ^{16}O HVA during exposure to $^{18}O_2$. The dotted line in the top graph represents concentration of HVA ($^{16}O + ^{18}O$). For full explanation see text.

V. POTENTIAL USEFULNESS OF IN VIVO LABELING OF ORGANIC COMPOUNDS BY EXPOSURE TO ^{18}O-CONTAINING ATMOSPHERES

The technique outlined in this chapter should be of value for the study of several problems of biological interest. The technique allows *in vivo* confirmation of *in vitro* data regarding hydroxylation mechanisms in a number of metabolic pathways consuming molecular oxygen. The present data also demonstrate the usefulness of *in vivo* labeling of organic compounds with ^{18}O for turnover studies. ^{18}O could thus be used to study turnover of dopamine and noradrenaline as well as 5-hydroxytryptamine in brain. Moreover, for studies on drug metabolism involving hydroxylation reactions the technique should be of interest. Exposure to ^{18}O appears to be a relatively nontoxic procedure (Samuel, 1973), which does not markedly alter physiological conditions in animals or man. The lack of reliable methods for determination of brain monoamine turnover in man makes the present application of ^{18}O of interest for clinical studies. Labeling of brain monoamines and subsequent determination of changes in specific activity of metabolites that can be detected in body fluids as cerebrospinal fluid, serum, or urine should allow calculation of turnover rates. This should be of value for future studies aimed at evaluating changes in brain monoamine synthesis in man during various neuropsychiatric disease states and to control specific drug therapies directed to alteration of brain monoamine metabolism. At the present time the main limitation of the procedure is the expense of isolating ^{18}O. With the increasing sensitivity of mass spectrometric detectors, which has constantly improved during the last few years, and the use of chemical ionization, it should be possible in the future to detect labeled monoamines in the human brain following a short time inhalation of $^{18}O_2$ gas. Future studies have to elucidate the practicability of such a method.

ACKNOWLEDGMENTS

The present study was supported by the Swedish Medical Research Council (Project No. B73 40X 3560 02A) and the Weizmann Institute of Science, Israel.

REFERENCES

Andén, N.-E., Roos, B.-E., and Werdinius, B. (1964): Effects of chlorpromazine, haloperidol and reserpine on the levels of phenolic acids in rabbit corpus striatum. *Life Sciences*, 3: 149–158.

Änggård, E., and Sedvall, G. (1969): Gas chromatography of catecholamine metabolites using electron capture detection and mass spectrometry. *Analytical Chemistry*, 41:1250–1256.

Costa, E. (1972): Appraisal of current methods to estimate the turnover rate of serotonin and catecholamines in human brain. *Advances in Biochemical Psychopharmacology, Vol. 4: Role of Vitamin B_6 in Neurobiology*, edited by M. S. Ebadi and E. Costa, pp. 171–183. Raven Press, New York.

Costa, E., and Groppetti, A. (1970): Biosynthesis and storage of catecholamines in tissues of rats injected with various doses of *d*-amphetamine. *International Symposium on Amphetamines and Related Compounds*, edited by E. Costa and S. Garattini, pp. 231–255. Raven Press, New York.

Daly, J., Levitt, M., Guroff, G., and Udenfriend, S. (1968): Isotope studies on the mechanism of action of acrenal tyrosine hydroxylase. *Archives of Biochemistry and Biophysics*, 126: 593–598.
Hornykiewicz, O. (1966): Dopamine (3-hydroxytyramine) and brain function. *Pharmacological Reviews*, 18:925–964.
Koslow, S. H., Cattabeni, F., and Costa, E. (1972): Norepinephrine and dopamine: Assay by mass fragmentography in the picomole range. *Science*, 176:177–180.
Mayevsky, A., Sjöquist, B., Fri, C. G., Samuel, D., and Sedvall, G. (1973): *Biochemical and Biophysical Research Communications*, 51:746–756.
Nagatsu, T., Levitt, M., and Udenfriend, S. (1964): The initial step in norepinephrine biosynthesis. *Journal of Biological Chemistry*, 239:2910–2917.
Neff, N. H., and Costa, E. (1967): Effect of tricyclic antidepressants and chlorpromazine on brain catecholamine synthesis. *Proceedings of the First International Symposium on Antidepressant Drugs*, pp. 28–34. Excerpta Medica Foundation, Amsterdam.
Nybäck, H., and Sedvall, G. (1968): Effect of chlorpromazine on accumulation and disappearance of catecholamines formed from ^{14}C-tyrosine in brain. *Journal of Pharmacology and Experimental Therapeutics*, 162:294–301.
Nybäck, H., and Sedvall, G. (1971): Effect of nigral lesion on chlorpromazine induced acceleration of dopamine synthesis from ^{14}C-tyrosine. *Journal of Pharmacy and Pharmacology*, 23:322–326.
Papeschi, R. (1972): Dopamine, extrapyramidal system, and psychomotor function. *Psychiatrica, Neurologia, Neurochirurgia*, 75:13–48.
Randrup, A., and Munkvad, I. (1967): Brain dopamine and the amphetamine-reserpine interaction. *Journal of Pharmacology and Pharmacy*, 19:483–484.
Samuel, D. (1962): Methodology of oxygen isotopes. In: *Oxygenases*, edited by O. Hayaishi, Academic Press, New York.
Samuel, D. (1973): Unpublished observations.
Shiman, R., Akino, M., and Kaufman, S. (1971): Solubilization and partial purification of tyrosine hydroxylase from bovine adrenal medulla. *Journal of Biological Chemistry*, 246:1330–1340.
Sjöquist, B., and Änggård, E. (1972): Determination of homovanillic acid in cerebrospinal fluid by gas chromatography with electron capture or mass spectrometric detection. *Analytical Chemistry*, 44:2297–2301.
Sjöquist, B., Dailey, J., Sedvall, G., and Änggård, E. (1973): Mass fragmentographic assay of homovanillic acid in brain tissue. *Journal of Neurochemistry*, 20:729–733.
Smith, T. E., Weissbach, H., and Udenfriend, S. (1962): Studies on the mechanism of action of monoamine oxidase: Metabolism of N,N-dimethyltryptamine and N,N-dimethyltryptamine-N-oxide. *Biochemistry Journal*, 1:137–143.
Sweeley, C. C., Elliot, W. H., Fories, J., and Ryhage, R. (1966): Mass spectrometric determination of unresolved components in gas chromatographic effluents. *Analytical Chemistry*, 38:1549–1553.
Wiesel, F.-A., Fri, C.-G., and Sedvall, G. (1973): Determination of homovanillic acid turnover in rat striatum by mass fragmentography and the use of monoamine oxidase inhibitors. *European Journal of Pharmacology, in press.*

Advances in Biochemical Psychopharmacology, Vol. 7
Raven Press, New York © 1973

The Use of Gas Chromatography-Mass Spectrometry to Measure Tissue Levels and Turnover of Acetylcholine

Donald J. Jenden

Department of Pharmacology, UCLA School of Medicine, Los Angeles, California 90024

One of the basic limitations of gas chromatographic analysis is that the compound which is chromatographed must have a significant vapor pressure at the temperature at which it is feasible to conduct the analysis. Quaternary ammonium compounds such as acetylcholine do not meet this requirement, and gas chromatographic analysis depends upon their reproducible and preferably quantitative conversion to a sufficiently volatile derivative. Acetylcholine has been derivatized in several ways, including reductive cleavage of the ester group to form ethanol (Stavinoha, Ryan, and Treat, 1964; Stavinoha and Ryan, 1965), hydrolysis to yield acetic acid (Cranmer, 1968), and N-demethylation (Szilagyi, Schmidt, and Green, 1968; Schmidt, Szilagyi, and Green, 1969; Jenden, Hanin, and Lamb, 1968; Hanin and Jenden, 1969). The last alternative has a significant advantage in that the derivative is unique, and is unlikely to be produced from other compounds present in a biological sample, whereas ethanol is a common trace contaminant and acetate is a universal constituent of biological systems. Two methods have been described for analysis of acetylcholine and related compounds as the N-desmethyl derivatives: pyrolysis of halide salts (Szilagyi et al., 1968; Schmidt et al., 1969) and reaction with sodium benzenethiolate (Jenden et al., 1968; Hanin and Jenden, 1969). The latter is the basis of the work described in the remainder of this chapter.

The reaction with sodium benzenethiolate (Fig. 1) has a high specificity for the methyl group when carried out in an aprotic solvent (Shamma, Deno, and Remar, 1966), and there is no significant attack on the ester group if strictly anhydrous conditions are maintained (Jenden et al., 1968), although under different conditions thiolate ions cleave esters (Sheehan and Daves, 1964; Vaughan and Baumann, 1962). N-Demethylation by sodium benzene-

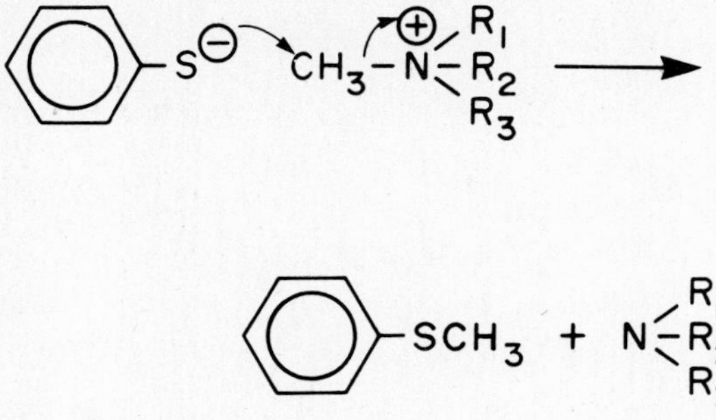

FIG. 1. N-Demethylation of quaternary ammonium compounds by nucleophilic displacement of a tertiary amine on a methyl group by benzenethiolate.

thiolate in butanone is quantitative at 80°C for 45 min. All choline esters which have been investigated undergo the same reaction, and it can also be used for alkyltrimethylammonium compounds and even for deethylation of hydroxyethyltriethylammonium and its esters (Jenden, *unpublished observations*).

In analyzing tissues for choline esters it is essential to employ a procedure in which the enzymatic systems that catalyze the synthesis and hydrolysis of acetylcholine are rapidly inactivated (Jenden and Campbell, 1971). In the work described here, tissues were rapidly removed and frozen by immersion in liquid nitrogen. The frozen tissues were pulverized in a stainless steel mortar and extracted with 15% 1N aqueous formic acid in acetone. A precisely known quantity of an internal standard is mixed with the homogenate, which is allowed to stand for 30 min before centrifugation, to release bound forms of acetylcholine. The supernatant is extracted twice with ether to remove lipids and most of the acetone; the remaining organic solvents are removed by evaporation with a stream of dry nitrogen. The remainder of the working procedure is summarized in Fig. 2.

If the starting sample is already in aqueous solution, the analysis is started by addition of the internal standard as in Fig. 2, and initial freezing is unnecessary. Quaternary ammonium compounds are precipitated by the addition of an equal volume of ammonium reineckate solution saturated at 0°C, after adding tetraethylammonium bromide as a coprecipitant to make a final concentration of 100 μM. After 45 min at 0°C, the suspension is centrifuged, the supernatant is discarded, and the precipitate is dried under vacuum.

Reineckate interferes with the demethylation reaction and can conveniently be removed by triturating the dry precipitate with silver *p*-toluenesulfonate (5 mM in acetonitrile), yielding a compact precipitate of silver

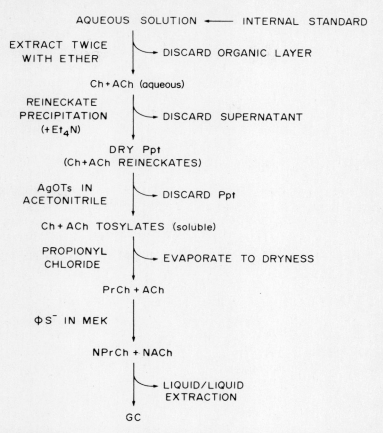

FIG. 2. Flow diagram for sample preparation in the analysis of aqueous samples for choline and acetylcholine.

reineckate, and leaving the quaternary ammonium compounds in solution as the *p*-toluenesulfonate salts (Jenden, Booth, and Roch, 1972). After centrifugation, the supernatant is transferred to a fresh centrifuge tube. If choline is to be estimated simultaneously, it is esterified at this stage for convenience in subsequent analysis by the addition of propionyl chloride. The reaction is complete in 5 min at room temperature, after which the solution is taken to dryness in a stream of dry nitrogen. The demethylation reagent is added and, after displacement of air from the reaction tube with nitrogen, the tube is capped and heated to 80°C for 45 min. After cooling, the solution is partitioned between aqueous acid and pentane. Tertiary amines are left in the aqueous phase while benzenethiol and methyl phenyl sulfide are removed in the organic phase. Finally, the aqueous solution is made basic (pH ~ 9.3) and the tertiary amines are extracted into a small volume of methylene chloride, an aliquot of which is injected into the gas chromatograph.

When a flame ionization detector is used, it is convenient to use trimethylacetylcholine as an internal standard. The limit of reliable detection is approximately 10 ng. Quantitation depends upon the linear relationship between the mole ratios of choline and acetylcholine to trimethylacetylcholine (Fig. 3), which are sequentially eluted from the gas chromatographic column, and on the exact quantity of internal standard initially added to the sample.

When used on freshly frozen brain, a peak was obtained at a retention time corresponding to N-demethylated acetylcholine, the height of which corresponded to the quantity of acetylcholine estimated by bioassay on the frog rectus abdominis (Hanin and Jenden, 1969). However, this does not constitute rigorous proof of the identity of the peak; for example, acetyl-β-

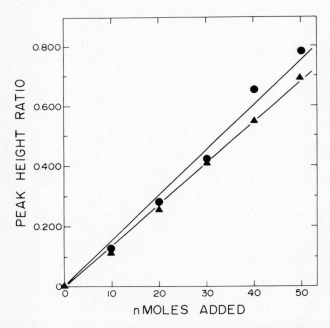

FIG. 3. Standardization curves for choline (▲) and acetylcholine (●) relating ratios of peak heights relative to the internal standard (trimethylacetylcholine) to mole ratio. Each analysis started with 1 ml aqueous solution containing 50 nmole internal standard. Of the 50 μl methylene dichloride in the final extract, 5 μl was injected into the gas chromatograph.

methyl choline gave a peak at an almost identical retention time under the gas chromatographic conditions employed (Fig. 4).

The peak in rat brain extracts at the appropriate retention time was shown to be acetylcholine by using the LKB 9000 gas chromatograph-mass spectrometer system equipped with an accelerating voltage alternator. The mass spectrum of the gas chromatographic peak from rat brain was the same as that obtained from acetylcholine carried through the entire procedure (Hammar, Hanin, Holmstedt, Kitz, Jenden, and Karlén, 1968) (Fig. 5).

The base peak of the spectrum is at *m/e* 58, and is due to the dimethyl-methylenimmonium ion (Hammar et al., 1968; Johnston, Triffett, and Wunderlich, 1968). By focusing the mass spectrometer on this mass, both the internal standard (in this case hexyltrimethylammonium) and acetylcholine appear in chromatograms of rat brain extracts. There are no peaks corresponding to propionylcholine or butyrylcholine, even though the gain

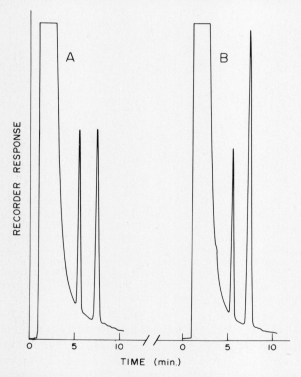

FIG. 4. Gas chromatograms of (A) hexyltrimethylammonium (1.5 nmole) and acetylcholine (3 nmole); and (B) hexyltrimethylammonium (1.5 nmole), acetylcholine (3 nmole), and methacholine (3 nmole) carried through the entire procedure.

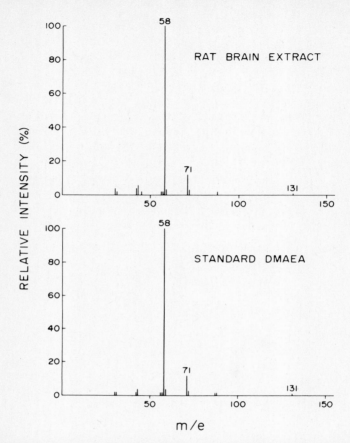

FIG. 5. Mass spectra obtained on LKB 9000 for dimethylaminoethyl acetate (DMAEA). Upper panel: demethylated brain extract. Lower panel: synthetic DMAEA. Gas chromatographic conditions: silanized glass column, 3 mm OD, 1 · 2 m; packed with 75% "Polypak I," 25% silanized "Gas Chrom P," coated with 1% dodecyldiethylenetriamine succinamide. Column temperature 128°C, injection port temperature 195°C. Carrier gas: helium at 3.5 kg/cm², and 25 ml/min. Mass spectrometric conditions: separator temperature, 200°C; ion source temperature, 310°C; ionization potential, 70 eV; multiplier voltage, 2 · 1 kV (Hammar et al., 1968).

was increased ×250, and it can be concluded that these esters do not occur in freshly frozen rat brain in significant amounts (< 5 pmoles). This allows the use of propionylcholine as a derivative for the estimation of choline, as described above.

For the remaining work, an EAI Quad 300 quadrupole mass spectrometer was coupled to a Varian 1400 gas chromatograph by means of a glass frit separator mounted in the detector oven. The quadrupole was programmed by a multiplexing device (Jenden and Silverman, 1971) which allows the instrument to be used to monitor the ion current at up to eight preselected masses. Separate outputs are provided for each mass, and may be recorded on a multichannel potentiometric recorder or separately integrated over a gas chromatographic peak.

This instrument has been used to identify acetylcholine definitively as the transmitter in the ciliary ganglion of the pigeon, a structure weighing approximately 300 μg and containing only 30 to 40 ng of the transmitter (Pilar, Jenden, and Campbell, 1973). The data in Table 1 were obtained by integrating the ion current at six of the important masses in the acetylcholine spectrum and comparing the relative abundance with that in the corresponding gas chromatographic peak from the ciliary ganglion.

TABLE 1. *Comparison of relative abundances of major fragment ions of the demethylation product of authentic acetylcholine with the component from a ciliary ganglion at an identical retention time*

m/e	Percent Relative Abundance	
	Authentic	Ganglion
42	46.7	46.6
43	24.1	26.0
58	100.0	100.0
71	7.5	7.0
72	1.5	2.1
131	< 1.0	< 1.0

Quantitative measurements by gas chromatography-mass spectrometry require the use of an internal standard as in conventional gas chromatography. An internal standard should ideally be as similar as possible to the compound to be measured in both physical and chemical characteristics in order to equalize losses; but it must be capable of separate measurement. Gaffney, Hammar, Holmstedt, and McMahon (1971) have pointed out that isotopically labeled variants of the compound to be analyzed approach this ideal when mass spectrometric detection is available, and in the present work two isotopic variants of choline and acetylcholine have been used (Jenden, Roch, and Booth, 1973). In all cases the base peaks of the derivatized compounds are due to the dimethylmethylenimmonium ion, which differs in mass depending on the site of deuterium labeling (Fig. 6). The D_4-

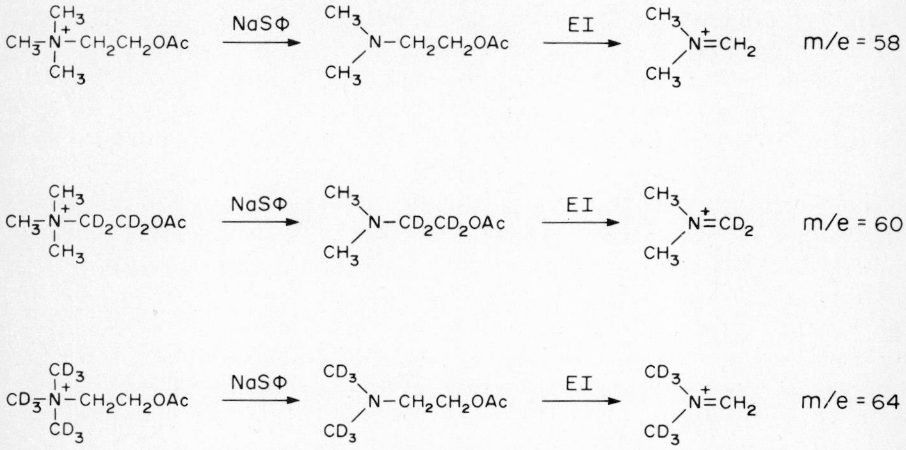

FIG. 6. Generation of base peaks at m/e 58, 60, and 64 by endogenous and deuterium-labeled variants of choline esters after benzenethiolate-induced N-demethylation and electron-impact (EI) ionization.

labeled compounds give base peaks at m/e 60 and the D_9-labeled variants give base peaks at m/e 64. The latter have been used as internal standards for the quantitative measurement of choline and acetylcholine (Jenden, 1972). By programming the mass spectrometer to monitor m/e 58 and 64, very accurate quantitation is possible to a limit of approximately 10^{-13} moles, based on the isotopic composition of single chromatographic peaks. Figure 7 is a representative chromatogram with the instrument operating in this mode, showing acetylcholine and choline peaks. Both have a 10:1 mole ratio of the D_9-labeled internal standards to unlabeled compounds (10 pmoles and 1 pmole respectively). Table 2 presents standardization data from which mole ratios may be accurately inferred from ion current ratios by linear interpolation.

It is convenient to use the D_4-labeled compounds for tracer studies, since by monitoring m/e 58, 60, and 64 absolute measurements may simultaneously be made of both the endogenous (D_0) and tracer (D_4) choline and acetylcholine by reference to D_9-labeled internal standards. Figure 8 presents this type of analysis of a sample from an experiment of the kind described by Schuberth, Sparf and Sundwall (1969). A pulse intravenous injection of D_4-choline (20 μmoles/kg) was given to a mouse, which was decapitated 1 min later. The brain was rapidly removed, frozen, pulverized, and homogenized as described above, using D_9-labeled internal standards. From the peak heights in Fig. 8 (allowing for the different sensitivity settings

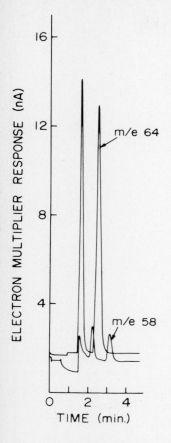

FIG. 7. Gas chromatogram with specific ion detection at m/e 58 and 64. Acetylcholine and choline were carried through the entire procedure from dilute aqueous solution. The amounts finally injected were 10 pmole (D_9-ACh and D_9-Ch) and 1 pmole (D_0-ACh and D_0-Ch).

of the three channels), it is possible to calculate that the quantities of endogenous choline and acetylcholine were 39.2 and 13.5 nmoles/g respectively, whereas the quantities of tracer (D_4) choline and acetylcholine were 0.519 and 0.337 nmole/g, respectively. The specific activities may be directly calculated as mole ratios: 0.0131 for choline and 0.0243 for acetylcholine.

From experiments of this sort in which measurements of endogenous and tracer choline and acetylcholine are made in brain at various times after the injection of tracer, it is possible to arrive at quantitative assessments of the turnover rate of acetylcholine. Similar experiments employing radioactive tracers generally require independent methods to establish total levels of choline and acetylcholine, without which measurements of radioactivity are of little value. Stable isotopic tracers also open the way to

TABLE 2. *Typical standardization data relating ion current ratio at m/e 58 and 64 to mole ratio of unlabeled, endogenous acetylcholine to internal standard. Calculated values correspond to a straight regression line*

Mole Ratio D_0/D_9 ACh	Ion current ratio m/e 58, 64	
	Observed	Calculated
0.000	0.022 0.023	0.020
0.200	0.273 0.273	0.274
0.400	0.521 0.522	0.528
0.600	0.777 0.777	0.782
0.800	1.063 1.040	1.036
1.000	1.280 1.285	1.290

multiple labeling experiments in which repeated pulse labels, separately measurable, allow the establishment of a kinetic picture in a single animal or experiment, since labels with ^2H, ^{13}C, and ^{15}N in different sites of the molecule may in general be separately identified and measured as a result of specific and predictable changes in the mass spectra of the precursor and products.

A series of experiments has been conducted in which these measurements of endogenous and tracer acetylcholine and choline have been made in brain at intervals of 20 sec to 30 min following pulse intravenous labeling. No significant change in the concentrations of acetylcholine and choline occurs following 20 μmoles/kg D_4-choline. From these data the average turnover rate has been estimated to be approximately 5 nmoles/g/min, in agreement with the conclusions of Schuberth, Sparf, and Sundwall (1970) using radioisotopes. Further experiments will be required to establish the functional pools of choline and acetylcholine in neural tissue and their exchange kinetics, the regional differences in kinetics which exist in brain, and the specific effects of various drugs. The techniques described appear well suited to these lines of investigation.

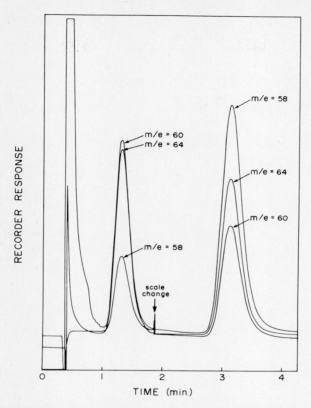

FIG. 8. Gas chromatogram of mouse brain with specific ion detection at m/e 58, 60, and 64 as indicated. The first set of peaks corresponds to acetylcholine, converted to dimethylaminoethyl acetate, with sensitivities of 100, 10, and 500 nA full scale for m/e 58, 60, and 64, respectively. The second set of peaks was due to choline, derivatized to dimethylaminopropionate, with sensitivities of 500, 50, and 500 nA for m/e 58, 60, and 64, respectively. Gas chromatographic conditions: silanized glass column 2 m × 2 mm I.D. packed with 5% DDTS, 5% OV 101 on GasChrom Q. Column temperature 98°C, injection port 150°C, helium flow rate 45 ml/min. Mass spectrometric conditions: separator temperature 150°C, electron energy 70 mV, emission current 250 μA, electron multiplier voltage 2.8 kV.

ACKNOWLEDGMENTS

The author is greatly indebted to Professor Bo Holmstedt for introducing him to the powerful and elegant technique of gas chromatography-mass spectrometry. The collaboration and assistance of many co-workers is gratefully acknowledged, including Drs. Israel Hanin and Carl-Gustav Hammar; and Robert W. Silverman, Margareth Roch, and Ruth A. Booth.

The research described was supported by grant MH 17691 from the National Institute of Mental Health.

REFERENCES

Cranmer, M. F. (1968): Estimation of the acetylcholine levels in brain tissue by gas chromatography of acetic acid. *Life Science,* 7:995–1000.

Gaffney, T. E., Hammar, C.-G., Holmstedt, B., and McMahon, R. E. (1971): Ion specific detection of internal standards labelled with stable isotopes. *Analytical Chemistry,* 43:307–310.

Hammar, C.-G., Hanin, I., Holmstedt, B., Kitz, R. J., Jenden, D. J., and Karlén, B. (1968): Identification of acetylcholine in fresh rat brain by combined gas chromatography/mass spectrometry. *Nature,* 220:915–917.

Hanin, I., and Jenden, D. J. (1969): Estimation of choline esters in brain by a new gas chromatographic procedure. *Biochemical Pharmacology,* 18:837–845.

Jenden, D. J. (1972): Simultaneous microestimation of choline and acetylcholine by gas chromatography/mass spectrometry/isotope dilution. *Federation Proceedings,* 31:515.

Jenden, D. J., Booth, R. A., and Roch, M. (1972): Simultaneous microestimation of choline and acetylcholine by gas chromatography. *Analytical Chemistry,* 44:1879–1881.

Jenden, D. J., and Campbell, L. B. (1971): Measurement of choline esters. In: *Analysis of Biogenic Amines and Their Related Enzymes,* edited by D. Glick, pp. 183–216. Interscience Publications, New York.

Jenden, D. J., Hanin, I., and Lamb, S. I. (1968): Gas chromatographic microestimation of acetylcholine and related compounds. *Analytical Chemistry,* 40:125–128.

Jenden, D. J., Roch, M., and Booth, R. A. (1973): Simultaneous measurement of endogenous and deuterium-labelled tracer variants of choline and acetylcholine in sub-picomole quantities by gas chromatography/mass spectrometry. *Analytical Biochemistry, in press.*

Jenden, D. J., and Silverman, R. W. (1971): An analog multiplexer for multiple specific ion detection in a gas chromatograph/quadrupole mass spectrometer system. In: *Proceedings of the Meeting on the Use of Stable Isotopes in Clinical Pharmacology,* edited by P. D. Klein and L. J. Roth, pp. 273–290. Document No. CONF 711115, National Technical Information Services, Springfield, Va.

Johnston, G. A. R., Triffett, A. C. K., and Wunderlich, J. A. (1968): Identification and estimation of choline derivatives by mass spectrometry. *Analytical Chemistry,* 40:1837–1840.

Pilar, G., Jenden, D. J., and Campbell, L. B. (1973): Distribution of acetylcholine in the normal and denervated pigeon ciliary ganglion. *Brain Research,* 49:245–256.

Schmidt, D. E., Szilagyi, P. I. A., and Green, J. P. (1969): Identification of submicrogram quantities of onium compounds by pyrolysis gas chromatography. *Journal of Chromatographic Sciences,* 7:248–249.

Schuberth, J., Sparf, B., and Sundwall, A. (1969): A technique for the study of acetylcholine turnover in mouse brain *in vivo. Journal of Neurochemistry,* 16:695–700.

Schuberth, J., Sparf, B., and Sundwall, A. (1970): On the turnover of acetylcholine in nerve endings of mouse brain *in vivo. Journal of Neurochemistry,* 17:461–468.

Shamma, M., Deno, N. C., and Remar, J. F. (1966): The selective demethylation of quaternary salts. *Tetrahedron Letters,* 13:1375–1379.

Sheehan, J. C., and Daves, G. D. (1964): Facile alkyl-oxygen ester cleavage. *Journal of Organic Chemistry,* 29:2006–2008.

Stavinoha, W. B., and Ryan, L. C. (1965): Estimation of the acetylcholine content of rat brain by gas chromatography. *Journal of Pharmacology and Experimental Therapeutics,* 150:231–235.

Stavinoha, W. B., Ryan, L. C., and Treat, E. L. (1964): Estimation of acetylcholine by gas chromatography. *Life Science,* 3:689–693.

Szilagyi, P. I. A., Schmidt, D. E., and Green, J. P. (1968): Microanalytical determinations of acetylcholine, other choline esters and choline by pyrolysis-gas chromatography. *Analytical Chemistry*, 40:2009–2013.

Vaughan, W. R., and Baumann, J. B. (1962): Reactions of alkyl carboxylic esters with mercaptides. *Journal of Organic Chemistry*, 27:739–744.

Advances in Biochemical Psychopharmacology, Vol. 7
Raven Press, New York © 1973

Uses of Stable Isotope Labeling with Gas Chromatography-Mass Spectrometry in Research on Psychoactive Drugs

D. R. Knapp, T. E. Gaffney, and K. R. Compson

Departments of Pharmacology and Medicine, College of Medicine, Medical University of South Carolina, Charleston, South Carolina 29401, and A.E.I. Scientific Apparatus Ltd., Manchester, England

I. INTRODUCTION

Stable isotopes have recently become the subject of renewed interest to the psychopharmacologist and indeed the biological scientist in general. This has been largely due to increased availability of gas chromatography-mass spectrometry (GC-MS) and the increased availability of the isotopes. Particular interest has been aroused in the area of clinical pharmacology (Knapp and Gaffney, 1972; Klein and Roth, 1972). We present here a summary of some uses of stable isotopes in biochemical psychopharmacology.

The greatest advantage offered by stable isotope labels over their radioactive counterparts in biochemical psychopharmacology studies is the absence of problems associated with radiation. Absence of possible radiation damage to living systems is very important in human investigation, especially in the case of children or pregnant women. Absence of radiation hazard also makes the synthesis and handling of stable isotope-labeled compounds much simpler.

Stability is less of a problem since there is no decay or radiolytic decomposition. In some cases, most notably that of deuterium versus tritium, the smaller mass difference of the stable isotope in relationship to the most abundant natural isotope makes for a smaller kinetic isotope effect, which is sometimes a source of confusion. In the case of nitrogen and oxygen the stable isotopes, ^{15}N and ^{18}O or ^{17}O, are without competition since there are no radioactive isotopes of sufficiently long half-life to be useful as labels.

The greatest single disadvantage of stable isotopes is the lack of a simple, sensitive, inexpensive isotope-specific method of quantitation comparable

83

to liquid scintillation counting; the stable isotope techniques require more expensive instrumentation and are associated with greater analytical complexity. With radioactive labels there is a very convenient, inexpensive, and sensitive non-compound-specific means of detection, that of detecting the emitted radiation. Modern liquid scintillation counters easily detect labeled drugs in picogram amounts, assuming typical specific activities in the 1 to 100 μC millimole range. Radiation detectors lack specificity, however, since the amount of isotope present in the sample is measured without regard to its chemical or molecular form. In contrast, the mass spectrometric technique of mass fragmentography provides a molecularly specific detection method whose sensitivity rivals that of radioisotopes in favorable circumstances, e.g., chlorpromazine has been detected in picogram amounts (Hammar, Holmstedt, and Ryhage, 1968).

Stable isotopes are useful in pharmacology as tracer labels for drug distribution and metabolism studies. The first use of stable isotope tracers in a biological study was reported in 1934 (Hevesy and Hofer). The same general technique of administering a labeled compound and determining the stable isotope abundance in an isolated product is, with refinements, still in use today (Fürst and Jonsson, 1971). In more recent development of the technique, isotope mixtures are administered and the artificially created mass spectral isotope cluster is used as a qualitative label. Such isotopes may also be useful as internal standards of quantitative measurements with mass spectrometry or GC-MS. Finally, stable isotope labels can aid in the use of exact mass measurements to establish the empirical formula for a molecule or fragment observed by mass spectrometry.

II. STABLE ISOTOPE-LABELED COMPOUNDS AS INTERNAL STANDARDS FOR QUANTITATIVE GC DETERMINATION OF PSYCHOACTIVE DRUGS AND OTHER COMPOUNDS

The isotope dilution method for quantitative determination was introduced by Rittenberg and Foster (1940) using stable isotopes. They used deuterium-labeled fatty acids and ^{15}N-labeled amino acids as adducts in quantitative determinations of these compounds. The subsequently isolated, purified, and combusted compounds were then examined for their relative isotopic abundance using an isotope ratio mass spectrometer. This is the same principle which underlies the determination of specific activity in the radioactive isotope dilution method.

The availability of the combined GC-MS instrument has made this technique applicable to intact molecules, i.e., it has eliminated the necessity of the combustion step and the need for a separate isolation and purification step. Samuelsson, Hamberg, and Sweeley (1970) have suggested this technique for the quantitative analysis of a natural product, prostaglandin E_1. Gaffney, Hammar, Holmstedt, and McMahon (1971) have suggested the technique for the quantitative measurement of the drug nortriptyline in human plasma (Fig. 1). In these techniques the stable isotope-labeled internal standard was added to the crude sample which was subjected to gas chromatographic separation and mass spectrometric analysis. In the drug application the labeled drug was added prior to extraction, derivatization, and analysis. In the prostaglandin application the internal standard was added as a deuterated methoxime derivative, a less desirable approach since prior extraction losses and variation in derivative formation from the natural material could introduce errors. In both these applications multiple-ion detection (Sweeley, Elliot, Fries, and Ryhage, 1966) was used to determine the ratio of labeled to unlabeled compounds. The technique has

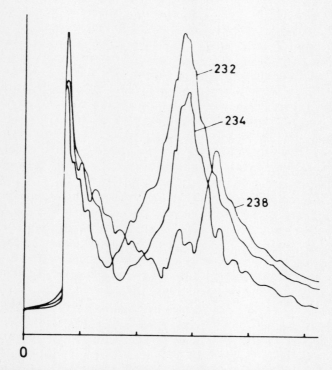

min   0

FIG. 1. Base peak mass fragmentograms of 50 pg each of trifluoroacetylated nortriptyline (mass 232), dideuterated nortriptyline (mass 234), and another standard IBD-18 (mass 238) (Gaffney et al., 1971).

subsequently been applied to other prostaglandins (Axen, Green, Horlin, and Samuelsson, 1971; Watson, Pelster, Dweetman, and Frolich, 1972), steroids (Siekmann, Hoppen, and Breuer, 1970), catecholamines (Koslow, Cattabeni, and Costa, 1972), and dimethyltryptamine (Vandenheuval, 1972).

This technique is associated with more complex instrumentation and more expense, and will therefore not replace older sensitive techniques such as electron capture detection (ECD) gas chromatography. For compounds not amenable to ECD sensitive derivatization (tertiary amines, for example) or for unstable and hard to isolate compounds, the technique has proven to be very useful.

III. USE OF STABLE ISOTOPE MIXTURES AS A QUALITATIVE TRACER TECHNIQUE FOR METABOLISM AND DISTRIBUTION STUDIES OF PSYCHOACTIVE DRUGS

Although mass spectrometry is the principal method for analysis of compounds labeled with stable isotopes, there is nothing diagnostic about the mass spectrum of the stable isotope-labeled compound except for the shift due to the increased mass. Unless the mass spectrum of the unlabeled compound is known, a labeled compound cannot be positively identified as such. Thus there is no way by mass spectral analysis alone to distinguish stable labeled compounds from other structures which are present in complex mixtures obtained from biological material. This problem can be overcome by the administration of mixtures of stable isotope-labeled and unlabeled compounds. The mixtures can be used to create a visually conspicuous isotope cluster in the mass spectrum.

Artificial isotopic doublets can thus serve as an easy means of identifying stable labeled compounds, either drugs or metabolites. Knapp, Gaffney, and McMahon (1972a) and Knapp, Gaffney, McMahon, and Kiplinger (1972b) have investigated the use of this technique in studies on the metabolism of the tricyclic antidepressant nortriptyline. They used M, M + 3 doublets created by mixing equimolar amounts of unlabeled drug and either trideuterium-labeled or dideuterium, ^{15}N-labeled drug. In their initial studies (Knapp, Gaffney, McMahon, and Kiplinger, 1971) they administered such a mixture to rats and identified a major metabolite in a complex urinary extract by observation of the M, M + 3 isotopic doublet in the mass spectrum of the compound (Fig. 2). Vore, Gerber, and Bush (1971) have reported on the same use of a M, M + 1 doublet created with a ^{15}N label in a barbiturate metabolism study. The M, M + 2 and M, M + 3 doublets seem to be preferable to M, M + 1 doublets, since natural isotopic contributions can con-

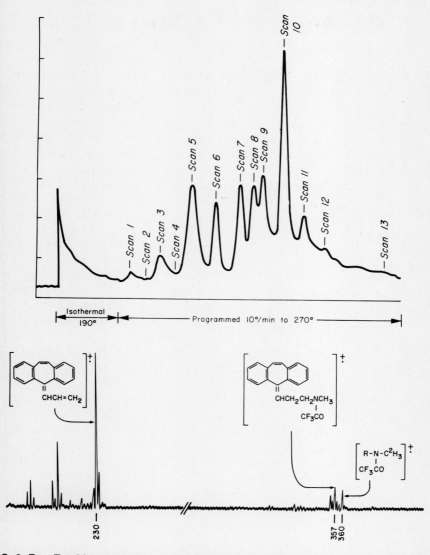

FIG. 2. *Top:* Total ion current recording from gas chromatographic separation of a basic urinary extract from a nortriptyline isotope mixture-dosed rat. *Bottom:* Partial mass spectrum from scan 10 showing *m/e* 357, 360 M, M + 3 doublet resulting from the mixture of stable isotope-labeled and -unlabeled compound (Knapp et al., 1972a).

tribute significantly to M + 1 peaks. This technique makes use of the mass spectrometer not only to identify the labeled chromatographic peaks in a complex chemical mixture but also to exploit the ability of the instrument to give information on the structure of the compounds and the location of the label.

An extension of this technique makes use of multiple-ion detection (Sweeley et al., 1966) to detect compounds containing stable isotope labels when the compounds are present in concentrations too low to be seen as a discrete total ion current peak and scanned, or when the compounds are obscured in the gas chromatograph trace by other peaks in the mixture. If the two masses of an equal intensity isotopic doublet are monitored by multiple-ion detection, one sees on the chromatographic trace two super-imposed peaks of equal intensity occurring at the retention time of the compound. Figure 3 shows results from a human urinary metabolite study in which subjects were given a single 25-mg dose of nortriptyline, half of which was isotopically labeled (Knapp et al., 1972b). In addition to the 10-hydroxy metabolite, which is visible as a peak in the total ion current recording, the metabolites desmethylnortriptyline and 10-hydroxydesmethylnortriptyline, as well as unchanged nortriptyline, were also observed. The identical technique was used to identify the same compounds in human bile after oral administration of only 25 mg of nortriptyline (Knapp et al., 1972b). These results indicate how the one-to-one intensity ratio of the two specific mass numbers adds a high degree of specificity to the sensitivity of the mass fragmentography technique. It is hoped that the use of this technique with chemical ionization mass spectrometry will yield a further increase in sensitivity.

Studies of drug disposition and drug action in the nervous system present the greatest challenge to the analytical techniques used in pharmacology. Further advances in these areas require the knowledge of the subcellular disposition of drugs within small, discrete areas of the brain. Localization and quantitation of drugs within such small areas as may be encountered when studying a single cell or a small group of cells require the utmost sensitivity and selectivity. Thus psychopharmacology research will be an impetus to development of and among the first to reap the benefits of improved analytical techniques such as those employing stable isotopes with GC-MS.

IV. USE OF STABLE ISOTOPE LABELS AS AN AID TO SELECTION OF CORRECT EMPIRICAL FORMULA FROM EXACT MASS MEASUREMENTS MADE IN LOW-RESOLUTION MODE

The sensitivity limitations of high-resolution mass spectrometers are presently such that to make exact mass measurements by electrical record-

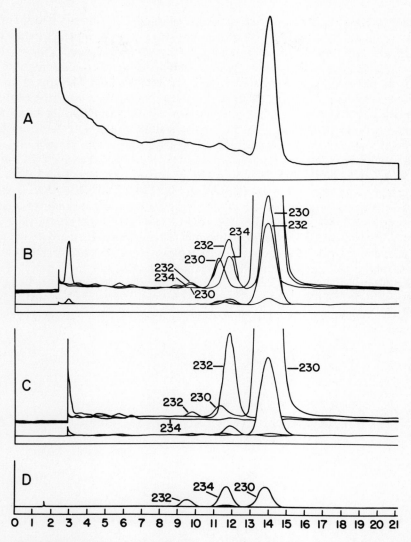

FIG. 3. Total ion current recording (A) and mass fragmentograms (B and C) of nortriptyline post-medication basic urine extract and reference compound mixture (D). Compounds in B have M, M + 2 base peak doublet. Ordinate scale is retention time in minutes
(Knapp et al., 1972b).

ing techniques on small quantities of material eluting from a gas chromatograph it is necessary to operate in low-resolution mode. This can be accom-
plished either by a peak-matching method on a single-beam instrument
(Hammar and Hessling, 1971) or by use of a dual-beam mass spectrometer.

Either method yields exact mass measurements that are less precise than those obtained by high-resolution mass spectrometry. As a result, when one calculates possible empirical formulas for the measured exact masses, one must allow wider error limits which results in an increased number of possible formulas from which to choose the correct one. The use of stable isotope labeling provides another tool for selecting from among the possible formulas. This use is a logical extension of the conventional use of natural abundance ^{13}C peaks as a check on the validity of a selected empirical formula. For example, if a selected empirical formula is a correct one, another peak of predictable relative intensity whose selected empirical formula has one ^{13}C in place of a ^{12}C should be observed one mass unit higher. In the case of mass measurements made on peaks from a stable isotope mixture experiment as described in the preceding section, one can look for a peak two or three mass units higher (depending upon the label employed) and check for a reasonable empirical formula for this peak that differs according to the label used. For example, if one had wished to make an exact mass measurement to get an empirical formula for the *m/e* 357 peak of the metabolite derivative's M, M + 3 doublet in Fig. 2, one could have used the fact that the selected empirical formula must have a corresponding empirical formula having three deuterium atoms in place of three hydrogens whose

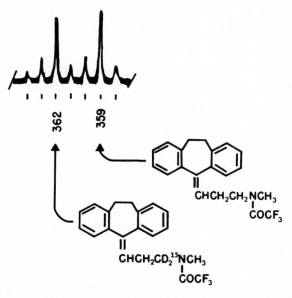

FIG. 4. Molecular ion doublet region from mass spectrum of trifluoroacetylated equimolar mixture of unlabeled and dideuterium, ^{15}N-labeled nortriptyline.

exact mass is within the error limits of the measured exact mass for the *m/e* 360 peak. In addition, ^{13}C natural abundance peaks could also be used in the conventional manner. These cumulative criteria of validity coupled with what is reasonable on a chemical basis would narrow the possible empirical formula choices to one or, at worst, a very few possibilities.

In actual experiments with an equimolar mixture of unlabeled and dideuterium, ^{15}N-labeled trifluoroacetyl nortriptyline, the exact masses of the peaks in the molecular ion doublet region (Fig. 4) were measured using a dual beam mass spectrometer (A.E.I. MS-30) operating at a nominal resolution of 1,000. A computer program then calculated possible empirical formulas with calculated exact masses within 25 ppm of the measured masses. The abbreviated table in Fig. 5 shows that 40, 40, 39, and 41

	Calculated Mass*	^{12}C	^{13}C	^{1}H	^{2}H	^{14}N	^{15}N	$\underline{O}$	$\underline{F}$	Measured Mass	Relative Intensity
40	363.1747	27	0	23	0	0	0	1	0	363.1697	0.45
	363.1625	20	1	18	2	0	1	1	3		
	363.1669	21	0	19	2	0	1	1	3		
40	362.1670	27	0	22	0	0	0	1	0	362.1622	1.55
	362.1591	21	0	18	2	0	1	1	3		
	362.1547	20	1	17	2	0	1	1	3		
39	360.1514	27	0	20	0	0	0	1	0	360.1514	0.55
	360.1530	20	1	20	0	1	0	1	3		
	360.1390	20	1	15	2	0	1	1	3		
41	359.1610	25	0	21	0	0	0	0	2	359.1534	1.65
	359.1496	21	0	20	0	1	0	1	3		
	359.1551	20	1	16	2	1	1	0	3		

* Allowable Error 25 ppm.

FIG. 5. Abbreviated computer generated table of possible empirical formulas for measured exact masses. Numbers on left are number of computed possibilities within error limits of 25 ppm for each measured mass. Of each group of three calculated formulas, first and last are the first and last of the list and the center is the selected correct formula.

possibilities, respectively, were calculated for the m/e 363, 362, 360, and 359 peaks. Only one formula of each of these groups fulfilled both the chemical (trifluoroacetyl derivative) and isotope label criteria of validity.

The ability to make exact mass measurements easily on gas chromatographic effluents (eliminating the need for subsequent isolation and purification) will be a boon to the biological scientist in his problems requiring structure proof. The employment of stable isotope techniques in conjunction with exact mass measurement promises to further increase this information yield.

ACKNOWLEDGMENT

Research supported in part by GM 20387–01.

REFERENCES

Axen, U., Green, K., Horlin, D., and Samuelsson, B. (1971): Mass spectrometric determination of picomole amounts of prostaglandins E_2 and $F_{2\alpha}$ using synthetic deuterium labelled carriers. *Biochemical and Biophysical Research Communications,* 45:519.

Fürst, P., and Jonsson, A. (1971): Control and modification of methods for determination of ^{15}N in biological material. *Acta Chemica Scandinavica,* 25:930–938.

Gaffney, T. E., Hammar, C.-G., Holmstedt, B., and McMahon, R. E., (1971): Ion specific detection of internal standards labeled with stable isotopes. *Analytical Chemistry,* 43:307–310.

Hammar, C.-G., and Hessling, R. (1971): Novel peak matching technique by means of a new and combined multiple ion detector-peak matcher device. *Analytical Chemistry,* 43:298–306.

Hammar, C.-G., Holmstedt, B., and Ryhage, R. (1968): Mass fragmentography: Identification of chlorpromazine and its metabolites in human blood by a new method. *Analytical Biochemistry,* 25:532–548.

Hevesy, G., and Hofer, E. (1934): How long does water remain in the body ("heavy" water as indicator). *Klinische Wochenschrift,* 13:1524.

Klein, P. D., and Roth, L. J., Eds. (1972): *Proceedings of a Seminar on the Use of Stable Isotopes in Clinical Pharmacology.* Publication number CONF-711115, National Technical Information Service, Springfield, Virginia.

Knapp, D. R., and Gaffney, T. E. (1972): Use of stable isotopes in pharmacology-clinical pharmacology. *Clinical Pharmacology and Therapeutics,* 13: 307–316.

Knapp, D. R., Gaffney, T. E., and McMahon, R. E. (1972a): Use of stable isotope mixtures as a labeling technique in drug metabolism studies. *Biochemical Pharmacology,* 21:425–429.

Knapp, D. R., Gaffney, T. E., McMahon, R. E., and Kiplinger, G. (1971): Studies on the metabolism of nortriptyline in the rat and man using stable isotope labeling. *Pharmacologist,* 13:220.

Knapp, D. R., Gaffney, T. E., McMahon, R. E., and Kiplinger, G. (1972b): Studies of human urinary and biliary metabolites of nortriptyline with stable isotope labeling. *Journal of Pharmacology and Experimental Therapeutics,* 180:784–790.

Koslow, S. H., Cattabeni, F., and Costa, E. (1972): Norepinephrine and dopamine: Assay by mass fragmentography in the picomole range. *Science,* 176:177–180.

Rittenberg, D., and Foster, G. L. (1940): A new procedure for quantitative analysis by isotope dilution with application to the determination of amino acids and fatty acids. *Journal of Biological Chemistry,* 133:737–744.

Samuelsson, B., Hamberg, M., and Sweeley, C. C. (1970): Quantitative gas chromatography of prostaglandin E_1 at the nanogram level. *Analytical Biochemistry,* 38:301–304.

Siekmann, L., Hoppen, H.-O., and Breuer, H. (1970): Determination of steroid hormones in body fluids by gas chromatography-mass spectrometry using a multiple ion detector. *Zeitschrift für Analytische Chemie*, 252:294–298.

Sweeley, C. C., Elliot, W. H., Fries, I., and Ryhage, R. (1966): Mass spectrometric determination of unresolved components in gas chromatographic effluents. *Analytical Chemistry*, 38:1549–1553.

Vandenheuvel, W. J. A. (1972): Application of GC techniques in drug metabolism and biochemistry. Presented at 4th Ohio Valley Gas Chromatography Symposium, Hueston Woods, Ohio, June 23–24, 1972.

Vore, M., Gerber, N., and Bush, M. T. (1971): The metabolic fate of 1-n-butyl-5,5-diethylbarbituric acid in the rat. *Pharmacologist*, 13:220.

Watson, J. T., Pelster, D., Dweetman, B. J., and Frolich, J. C. (1972): Prostaglandin analysis with a GC-MS-computer system. Presented at the 20th Annual Conference on Mass Spectrometry and Allied Topics, June 4–9, 1972, Dallas, Texas.

Advances in Biochemical Psychopharmacology, Vol. 7
Raven Press, New York © 1973

Studies on the Disposition of Amitriptyline and Other Tricyclic Antidepressant Drugs in Man as It Relates to the Management of the Overdosed Patient

H. Gard, D. Knapp, I. Hanenson, T. Walle, and T. Gaffney

Departments of Pharmacology and Medicine, Medical University of South Carolina, Charleston, South Carolina 29401, and Division of Clinical Pharmacology, Departments of Pharmacology and Internal Medicine, University of Cincinnati College of Medicine, Cincinnati, Ohio 45219

I. INTRODUCTION

Despite the widespread use and recognized hazards of amitriptyline therapy (Rasmussen, 1965; Ward and Tin-Myint, 1965; Moir, Cornwell, Dingwall-Fordyce, Crooks, O'Malley, Turnbull, and Weir, 1972), there is scant quantitative data on the disposition of this drug and its metabolites in man. For example, there are no data on the amounts of this drug and its metabolites which are secreted into gastric or biliary juices in man. This dearth of information exists even though animal studies with similar compounds (McMahon, Marshall, Culp, and Miller, 1963) and the clinical course of some overdose victims suggest that enterohepatic recycling of the drugs and of their metabolites could contribute to the development of delayed toxicity. Sequential, quantitative data on the disposition of the drug and its metabolites are also unavailable in man even though it is recognized that the onset of the most severe cardiac toxicity may be delayed for 24 hr or more after ingestion of an overdose and unexpected death has been reported to occur as late as 4 days after drug ingestion (Sedal, Korman, Williams, and Mushin, 1972).

The possible importance of quantitative studies on the metabolism and disposition of amitriptyline in man is illustrated by reports on the blood, urine, and tissue levels of parent drug and metabolites in patients receiving therapeutic doses (Braithwaite and Widdop, 1971) and in overdose victims (Munksgaard, 1969; Breyer and Remmer, 1971) (see Table 1). Considering the levels of metabolites found in these body fluids and tissues, the question is whether the observed toxicity resulted from the parent drug, its metabolites, or both. The presence of large amounts of pharmacologically active metabolites such as nortriptyline within these tissues may be explained by secretion of the metabolite from the liver into the blood, enterohepatic cycling, or both. Extrahepatic metabolism could also contribute to the tissue burden of some of these metabolites (Whitnack, Knapp, Holmes, Fowler, and Gaffney, 1972).

TABLE 1. *Drug and metabolite concentrations observed in fatal amitriptyline overdose cases (mg/ml)*

| | Amitriptyline | Nortriptyline | Unidentified metabolites | | |
			A	B	C
Case ONE*					
(24 to 36 hr postingestion)					
Blood	6	5	4	—	—
Urine	6	10	—	34	trace
Case TWO**					
(48 hr postingestion)					
Liver	50	51	—	—	—
Lung	46	78	—	—	—
Kidney	26	21	—	—	—
Heart muscle	12	14	—	—	—

*Munksgaard, 1969.
**Breyer and Remmer, 1971.

The lack of sequential, quantitative data on the disposition of amitriptyline and its metabolites in overdose victims has led us to make sequential measurements of the amounts of amitriptyline, nortriptyline, 10-hydroxy-amitriptyline, and 10-hydroxynortriptyline in gastric juice, urine, and plasma of a patient who had received a known high amount of amitriptyline while hospitalized; a similar study was done in a second patient. The biliary excretion of nortriptyline and its metabolites was estimated in a third patient who was given a single oral dose of 25 mg of nortriptyline.

Finally, because of the lack of rigorous documentation of the affinity of charcoal for amitriptyline and the common practice of administering charcoal to amitriptyline overdose victims, the degree and reversibility of amitriptyline adsorption from human gastric juice by charcoal was quantitatively examined under acidic and basic conditions.

II. RESULTS

A. Case One

R. F., a 23-year-old male, ingested 3,750 mg of amitriptyline while hospitalized in a psychiatric ward. The first gastric lavage 2.5 hr after ingestion removed 204 mg of amitriptyline and trace amounts of the metabolites nortriptyline, 10-hydroxyamitriptyline, and 10-hydroxynortriptyline. This initial lavage removed 5.4% of the dose. In the 24 hr after the initial lavage, continuous gastric aspiration removed an additional 4.5% of the dose as parent drug and metabolites.

Using an indwelling catheter, continuous urine collection yielded 3.4% of the ingested dose in 26.5 hr. A total of 13.3% of the ingested dose was accounted for by gastric lavage, gastric aspiration, and urinary excretion during the first 24 hr. Figure 1 shows typical gas chromatograph (GC) traces for parent drug and metabolites obtained during the analysis of blood, urine, and gastric juice: note the excellent peak symmetry which made derivatization unnecessary. The identity of each GC peak was confirmed by compari-

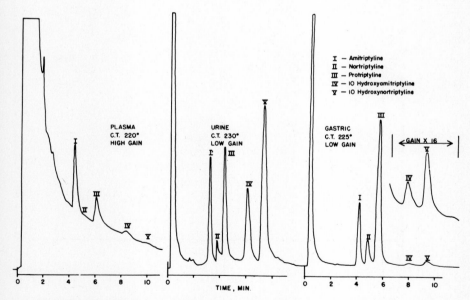

FIG. 1. Gas chromatograms of the hexane extracts of three human body fluids (at pH 14) using a flame ionization detector.

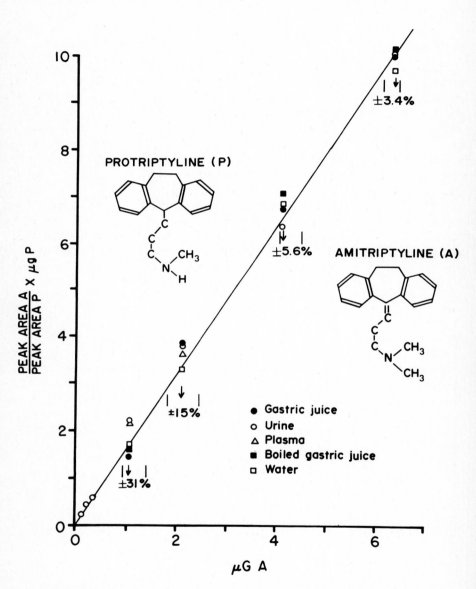

FIG. 2. Amitriptyline standard curve with protriptyline as the internal standard.

son of low-resolution mass spectra with mass spectra of reference compounds.

Estimating 5.5 liters of blood in this 70 kg subject, 3.2% of the ingested amitriptyline, or approximately 120 mg, was present in the blood 2.5 hr after ingestion. After removal through lavage, aspiration, and urinary excretion of 13.3% of the initial dose in the first 26.5 hr, the total blood volume still contained 2.1% of the initial dose. Figure 2 shows the standard curve for amitriptyline with protriptyline as the internal standard. The amounts of 10-hydroxyamitriptyline and 10-hydroxynortriptyline were estimated from the standard curves of the corresponding nonhydroxylated compounds.

B. Case Two

R. G., a 45-year-old female, ingested 500 mg of amitriptyline in a suicide attempt. The first gastric lavage 3 hr after ingestion removed 48.2 mg, or 9.6% of the drug. Subsequently, continuous gastric aspiration until 27 hr after ingestion removed an additional 16% of the original dose as parent drug and metabolites. Urinary excretion accounted for only 1.3% of the original dose in the first 27 hr. A total of 27% of ingested drug was removed in the first 27 hr. The plasma samples from this patient were lost.

C. Case Three

K. E., a 25-year-old female who had undergone elective cholecystectomy and had a bile duct "T" tube in place, volunteered for this study. A single 25-mg oral dose of nortriptyline was administered on the third postoperative day, and bile was collected for the following 24 hr. Since the "T" tube does not remove the total bile production, the estimated concentrations in the collected bile and an estimate of normal adult bile production of about 500 ml/day were used to estimate that approximately 2% of the dose was excreted in the bile as parent drug and about 13% as the 10-hydroxy-nortriptyline metabolite during the first 24 hours. This estimate may be low in the glucuronide fraction, because subsequent testing indicated that the glucuronidase was slightly degraded.

D. Charcoal-Amitriptyline Binding in Human Gastric Juice

In vitro studies showed that 1 g of charcoal binds 95% of 10 mg of amitriptyline in 10 ml of clear human gastric juice; 88% of the amitriptyline was absorbed when 21 mg of the drug was present (Fig. 3).

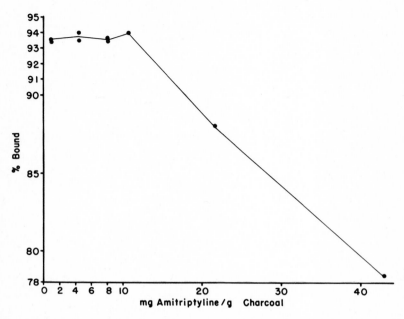

FIG. 3. Adsorption of amitriptyline from human gastric juice by activated charcoal.

III. DISCUSSION

It is apparent from these and earlier reported studies that the highly lipid-soluble drug amitriptyline and its metabolites are primarily localized in tissues. Hence, only small amounts of the drug and its metabolites are in the blood and a large percent of the blood content is protein bound (Borgå and Lunde, 1970). Binding to plasma proteins and localization in tissues reduce the amitriptyline filtration rate and its excretion by the kidney. Less than 3.2% appears in the urine during the first 24 hr after ingestion of a massive dose.

The gas chromatographic-mass spectrometric determination reveals that significant amounts of each of the major metabolites of amitriptyline are secreted into the gastric juice. This fate of amitriptyline has not been previously reported and emphasizes the complexities facing the pharmacokineticist in drug disposition studies in man. Our initial observations of this phenomenon were made after overnight fasts in two patients under treatment with therapeutic doses of imipramine and amitriptyline respectively. TFA derivatives of basic extracts of fasting morning specimens of gastric juice in these patients revealed the presence of imipramine, desmethylimipramine, and amitriptyline nortriptyline, 10-hydroxynortriptyline (Figs. 4–6).

Although the amounts of amitriptyline and amitriptyline equivalents of metabolites removed by lavage and continuous aspiration were not large, continuous gastric aspiration during the first 24 hr after initial gastric lavage removed amounts of metabolites and parent drug equivalent to or in excess of that removed by the first lavage shortly after drug ingestion. The total amount removed by lavage and continuous suction approaches 20% of the total ingested dose.

The gastric secretion of both psychopharmacologically active and other metabolites appears to provide for another type of drug recycling similar to enterohepatic recycling. The quantitative importance of this type of recycling mechanism remains to be determined.

Enterohepatic recycling of tricyclic antidepressant drugs and their metabolites also appears possible following biliary excretion. Knapp, Gaffney,

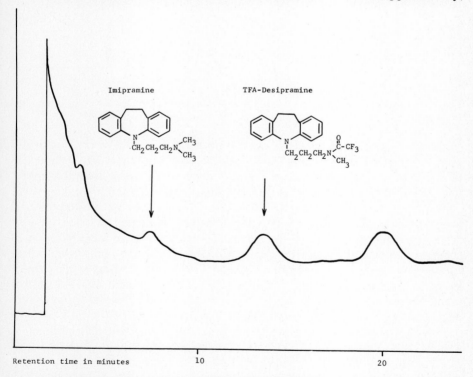

FIG. 4. The total ion current recording from the GC-MS examination of gastric juice extract from the patient who was given imipramine.

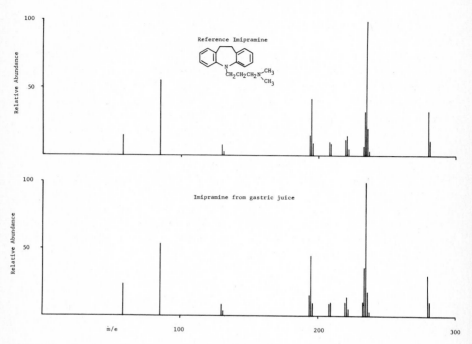

FIG. 5. Comparison of mass spectrum of reference imipramine (top) with that of imipramine found in gastric juice (bottom).

McMahon, and Kiplinger (1972) found nortriptyline, 10-hydroxynortripty-
line, desmethylnortriptyline, and 10-hydroxydesmethylnortriptyline in
human bile. In the present study we have observed in one patient that more
than 15% of an oral dose of nortriptyline was excreted in human bile
within the first 24 hr after a single 25-mg dose. If tricyclic antidepressant
drugs are regularly excreted into the human bile in this or even greater
amounts, reabsorption seems likely since analysis of human feces has shown
only small percentages of tricyclic antidepressant drugs to be present after
therapeutic doses (Eschenhof and Rieder, 1969). McMahon et al. (1963)
found the biliary excretion of nortriptyline to be as high as 55% of the
administered dose in rats in carefully performed studies. Similar results have
been reported for amitriptyline (Ryrfeldt and Hansson, 1971).

   The high tissue concentration of active metabolites of amitriptyline seen in
some overdose cases and the reports of unexplained and delayed death 3 to 4
days after drug ingestion (Sedal et al., 1972), coupled with the potential
for drug recycling from gastric and biliary sources, indicate the need for
more extensive quantitative studies of drug release into human gastric and

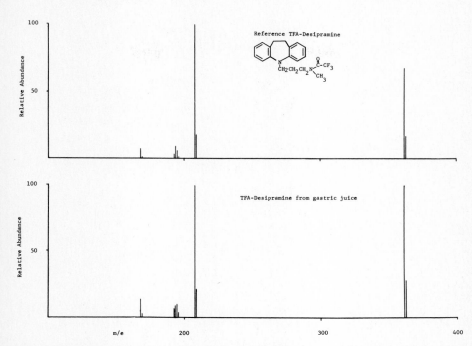

FIG. 6. Comparison of mass spectrum of reference trifluoroacetylated (TFA) desipramine (top) with TFA-desipramine in gastric juice (bottom).

biliary juices in overdosed patients. In any event, it appears from these preliminary observations that gastric and biliary secretion of tricyclic anti-depressant drugs and their metabolites might account for as much as 30% of the ingested dose during the first 24 hr (Fig. 7). These observations, coupled with the documentation of charcoal affinity for amitriptyline, suggest that after initial gastric lavage the management of tricyclic overdose cases should include installation of charcoal into the stomach and in the duodenum in the vicinity of the ampulla of Vater at intervals of perhaps 4 to 6 hr throughout the first 24 or 48 hr after ingestion.

It should be emphasized that many studies of the adsorptive power of charcoal for drugs have apparently been carried out in simple water solutions. These studies have yielded adsorptive ratios approaching unity for some drugs. Charcoal's capacity to bind amitriptyline in the present study was determined in one of the actual media from which amitriptyline must be removed, i.e., the gastric juice. In normal gastric juice much of the charcoal is immediately deactivated by natural chemical constituents. Consequently, 1 gm of charcoal was observed to bind 79% of 43 mgm of amitripty-

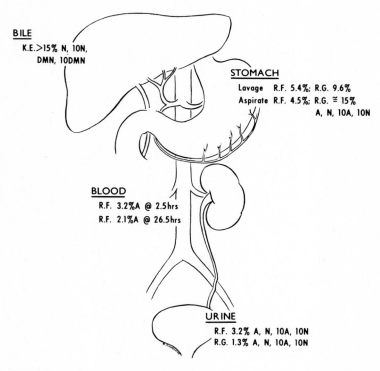

BILE
K.E.>15% N, 10N,
DMN, 10DMN

STOMACH
Lavage R.F. 5.4%; R.G. 9.6%
Aspirate R.F. 4.5%; R.G. ≅ 15%
A, N, 10A, 10N

BLOOD
R.F. 3.2%A @ 2.5hrs
R.F. 2.1%A @ 26.5hrs

URINE
R.F. 3.2% A, N, 10A, 10N
R.G. 1.3% A, N, 10A, 10N

FIG. 7. Drug-distribution results from three patients in this study showing percentage of ingested doses of tricyclic antidepressant drugs found in human gastric juice, bile, and urine during first 24 hr.

line. These findings suggest that an initial dose of 100 gm of charcoal could be expected to inactivate 4 gm of ingested amitriptyline.

ACKNOWLEDGMENT

This work was supported in part by GM 20387–01.

REFERENCES

Borgå, O., and Lunde, P.K.M. (1970): Protein binding of nortriptyline and diphenylhydantoin in man. *Acta Pharmacologica et Toxicologica*, 28:16–17.
Braithwaite, R. A., and Widdop, B. (1971): A specific gas-chromatographic method for the measurement of "steady-state" plasma levels of amitriptyline and nortriptyline in patients. *Clinica Chimica Acta*, 35:461–472.
Breyer, U., and Remmer, H. (1971): Determination of amitriptyline and metabolites in various organs after fatal poisoning. *Archiv Toxikologie*, 28:176–181.

Eschenhof, Von E. and Rieder, J. (1969): Studies on the fate of the antidepressant amitripty-line in the organisms of rat and man. *Arzneimittel-Forschung,* 19:957–966.

Knapp, D. R., Gaffney, T. E., McMahon, R. E., and Kiplinger, G. (1972): Studies of human urinary and biliary metabolites of nortriptyline with stable isotope labeling. *Journal of Pharmacology and Experimental Therapeutics,* 180:784–790.

McMahon, R. E., Marshall, F. J., Culp, H. W., and Miller, W. M. (1963): The metabolism of nortriptyline-N-methyl-^{14}C in rats. *Biochemical Pharmacology,* 12:1207–1217.

Moir, D. C., Cornwell, W. B., Dingwall-Fordyce, I., Crooks, J., O'Malley, K., Turnbull, M. J., and Weir, R. D. (1972): Cardiotoxicity of amitriptyline. *Lancet,* 2:561–564.

Munksgaard, E. C. (1969): Concentrations of amitriptyline and its metabolites in urine, blood and tissue in fatal amitriptyline poisoning. *Acta Pharmacologica et Toxicologica,* 27:129–134.

Rasmussen, J. (1965): Amitriptyline and imipramine poisoning. *Lancet,* 2:850–851.

Ryrfeldt, A., and Hansson, E. (1971): Biliary excretion of quaternary ammonium compounds and tertiary amines in the rat. *Acta Pharmacologica et Toxicologica,* 30:59–68.

Sedal, L., Korman, M. G., Williams, P. O., and Mushin, G. (1972): Overdosage of tricyclic antidepressants — A report of two deaths and a prospective study of 24 patients. *Medical Journal of Australia,* 2:74–79.

Ward, F. G., and Tin-Myint, B. (1965): Amitriptyline poisoning. *Lancet,* 2:910.

Whitnack, E., Knapp, D. R., Holmes, J. C., Fowler, N. O., and Gaffney, T. E. (1972): De-methylation of nortriptyline by the dog lung. *Journal of Pharmacology and Experimental Therapeutics,* 181:288–291.

Advances in Biochemical Psychopharmacology, Vol. 7
Raven Press, New York © 1973

The Psychopharmacological Effects of a New Metabolite of Propranolol

David A. Saelens, Thomas Walle, Philip J. Privitera, Daniel R. Knapp, and Thomas Gaffney

Departments of Pharmacology and Medicine, Medical University of South Carolina, Charleston, South Carolina 29401

I. INTRODUCTION

The mechanisms by which propranolol produces its effects on the central nervous system are poorly understood. Although propranolol is extensively metabolized (Bond, 1967), it is generally assumed that the effects on the central nervous system result from the action of unchanged propranolol. Recent studies in our laboratory on the metabolic fate of propranolol in man and dogs have led to the identification of 16 new metabolites of this compound by gas chromatography-mass spectrometry (GC-MS) (Walle, Ishizaki, Saelens, Privitera, Garteiz, and Gaffney, 1972; Walle and Gaffney, 1972*a,b;* Walle, Ishizaki, and Gaffney, 1973). One of the major metabolites identified was a glycerol ether, 3-(α-naphthoxy)-1,2-propanediol (propranolol glycol), with close structural similarity to the central muscle relaxant mephenesin (Fig. 1).

Propranolol and mephenesin have been reported to produce similar central effects in experimental animals, i.e., quieting, ataxia, and paralysis (Berger, 1949; Leszkovszky and Tardos, 1965; Murmann et al., 1966). Both compounds exhibit anticonvulsant (Berger, 1949; Leszkovszky and Tardos, 1965; Murmann et al., 1966) and internuncial neuron-blocking activity (Sinha, Srimal, Jajer, and Bhargava, 1967). Furthermore, both drugs have been advocated for the treatment of parkinsonism (Schlesinger, Drew, and Wood, 1948; Owen and Marsden, 1965; Abramsky, Carmon, and Lavy, 1971; Gilligan, Veale, and Wadak, 1972), and anxiety states (Schlan and Unna, 1949; Wheatly, 1969; Linken, 1971). More recently, propranolol has been reported to block the "high" produced by heroin (Grosz, 1972) and alcohol (Mendelson, Rossi, and Bernstein, 1972).

The similarity between propranolol and mephenesin in their effects on the central nervous system suggests that the newly discovered glycol metabolite

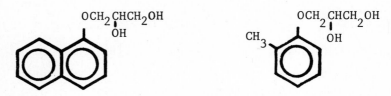

FIG. 1. Structural formulas for propranolol glycol (*left*) and mephenesin (*right*).

of propranolol might be contributing to some of the central effects of propranolol. The effects of propranolol and propranolol glycol on gross behavior and strychnine-induced convulsions were compared in mice.

II. BEHAVIORAL AND ANTICONVULSANT EFFECTS OF PROPRANOLOL AND PROPRANOLOL GLYCOL

Both i.v. propranolol and propranolol glycol produced quieting at low doses. Higher doses of propranolol glycol resulted in ataxia and paralysis (Table 1).

TABLE 1. *Incidence of paralysis and death after i.v. propranolol glycol*

Prooranolol glycol (mg/kg)	No. of animals tested	% of Animals	
		Paralyzed	Dead
20	5	0	0
40	10	100	0
80	12	100	17
95	9	100	89
120	10	100	90
160	3	100	100

TABLE 2. *Protective effect of propranolol glycol against death due to strychnine convulsions*

Propranolol glycol (mg/kg)	Strychnine sulfate 0.43 mg/kg	
	injected animals	surviving animals
0	10	1
5	5	0
10	10	6
20	10	7
40	10	10

Propranolol glycol reduced the lethal effect of strychnine-induced convulsions in mice (Table 2). Propranolol had the same effect. The anticonvulsant activity of propranolol was not maximal until 10 min after injection, whereas the protective effect of propranolol glycol was immediate. The duration of the anticonvulsant effect was less than 30 min for both compounds.

III. PRESENCE OF A DIOL METABOLITE IN MOUSE BRAIN AFTER PROPRANOLOL INJECTION

The 2,3-propanediol metabolite of propranolol was detectable in brains of mice sacrificed 10 min after the i.v. administration of propranolol. It was identified as its trifluoroacetyl derivative by GC-MS using mass fragmentography (Hammar, Holmstedt, and Ryhage, 1968).

Brains were homogenized in 0.4 N $HClO_4$ and centrifuged. The supernatant was extracted with benzene and the extract derivatized with trifluoroacetic anhydride prior to GC-MS. The mass spectrometer was focused at m/e 410, the base peak and molecular ion of the ditrifluoroacetyl derivative of the diol metabolite (Fig. 2).

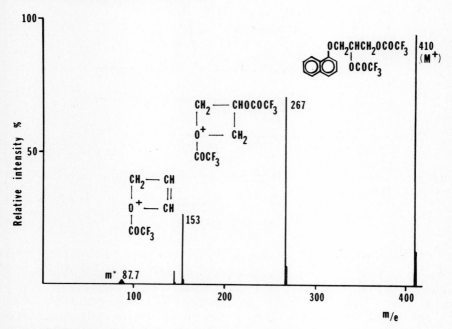

FIG. 2. Normalized mass spectrum of ditrifluoroacetylated reference compound propranolol glycol at 20 eV.

The resulting single-ion mass fragmentograms (Fig. 3) demonstrate a significant peak of m/e 410 in the brains only after propranolol administration. This peak had exactly the same retention time as that of the reference diol compound. There was no interference from either propranolol or other propranolol metabolites under these conditions.

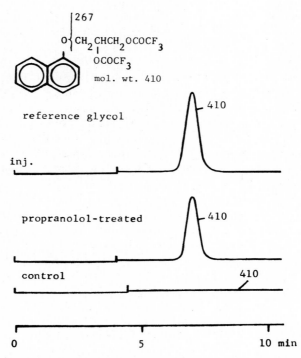

FIG. 3. GC-MS of trifluoroacetylated extracts of mouse brain before and after propranolol injection.

IV. DISCUSSION

Our data indicate that both propranolol and its diol metabolite, propranolol glycol, produce the same gross behavioral effects in mice. In addition, both compounds protected mice against the lethal effects of strychnine-induced convulsions. The onset of this anticonvulsant effect was immediate for the glycol metabolite, but was delayed after the administration of propranolol. These observations strongly suggest that the delay in onset of this anticonvulsant effect may be due to the time required for the formation of an active metabolite, i.e., propranolol glycol. The finding of propranolol glycol in brain tissue 10 min after injection of propranolol further supports this hypothesis.

The immediate onset of behavioral effects of propranolol excludes the possibility that the lag time for development of maximal anticonvulsant activity was related to slow uptake of propranolol into the central nervous system.

Since the biotransformation of propranolol to propranolol glycol resembles the conversion of catecholamines in brain to their respective glycol products (Walle and Gaffney, 1972b), it is possible that this glycol metabolite may be formed locally within the brain. Although it seems unlikely that propranolol itself would be a substrate for monoamine oxidase within brain tissue, the N-desisopropyl metabolite of propranolol (Walle and Gaffney, 1972b), which is a primary amine, is a substrate for monoamine oxidase (Walle and Gaffney, 1972a) and might be expected to be converted through an intermediate aldehyde to the diol metabolite within brain tissue. Since the same enzyme system which metabolizes catecholamines may be involved in the formation of propranolol glycol, it is possible that administration of propranolol may affect catecholamine metabolism within the central nervous system (Sullivan, Segal, Kuczenski, and Mandell, 1972).

Since propranolol and mephenesin are reported to have the same spectrum of activity in the central nervous system, our findings of an active diol metabolite of propranolol suggest that some of the central effects of propranolol may be related to the formation of propranolol glycol. Indeed, the formation of this active metabolite and possibly other metabolites not yet studied may explain why propranolol appears to have therapeutic effects in a variety of central nervous system disorders. The recently reported usefulness of propranolol in heroin addiction (Grosz, 1972) and alcoholism (Tyler, 1972) may also be related to the activity of formed metabolites rather than propranolol itself.

These findings with propranolol glycol suggest that at least some of the effects of propranolol result from the formation of active metabolites.

ACKNOWLEDGMENT

The research presented here was supported in part by U.S. Public Health Service grant GM 20387–01.

REFERENCES

Abramsky, O., Carmon, A., and Lavy, S. (1971): Combined treatment of parkinsonian tremor with propranolol and levodopa. *Journal of the Neurological Sciences*, 14:491–494.
Berger, F. M. (1949): Spinal cord depressant drugs. *Pharmacological Reviews*, 1:243–278.

Bond, P. A. (1967): Metabolism of propranolol ("Inderol"), a potent, specific β-adrenergic receptor blocking agent. *Nature,* 231:721.

Gilligan, B. S., Veale, J. L., and Wadak, J. (1972): Propranolol in the treatment of tremor. *Medical Journal of Australia,* 1:320–322.

Grosz, H. J. (1972): Narcotic withdrawal symptoms in heroin users treated with propranolol. *Lancet,* 2:564–566.

Hammar, C.-G., Holmstedt, B., and Ryhage, R. (1968): Mass fragmentography. Identification of chlorpromazine and its metabolites in human blood by a new method. *Analytical Biochemistry,* 25:532–548.

Leszkovszky, G., and Tardos, L. (1965): Some effects of propranolol on the central nervous system. *Journal of Pharmacy and Pharmacology,* 17:518–520.

Linken, A. (1971): Propranolol for L.S.D.-induced anxiety states. *Lancet,* 2:1039–1040.

Mendelson, J. H., Rossi, A. M., and Bernstein, J. (1972): Effects of propranolol on behavior of alcoholics following acute alcohol intake. *Abstracts of Volunteer Presentations, Fifth International Congress on Pharmacology,* p. 157. San Francisco, California.

Murmann, W., Almirante, L., and Saccani-Guelfi, M. (1966): Central nervous system effects of four β-adrenergic receptor blocking agents. *Journal of Pharmacy and Pharmacology,* 18:317–318.

Owen, D. A. and Marsden, C. D. (1965): Effect of adrenergic beta-blockade on parkinsonian tremor. *Lancet,* 2:1259–1262.

Schlan, L. S., and Unna, K. R. (1949): Some effects of myanesin in psychiatric patients. *Journal of American Medical Association,* 140:672–673.

Schlesinger, E. B., Drew, A. L., and Wood, B. (1948): Clinical studies in the use of myanesin. *American Journal of Medicine,* 4:365–372.

Sinha, J. N., Srimal, R. C., Jajer, B. P., and Bhargava, K. P. (1967): On the central muscle relaxant activity of D.C.I., nethalide and propranolol. *Archives Internationales De Pharmacodynamie Et De Therapie,* 165:160–166.

Sullivan, J. L., Segal, D. S., Kuczenski, R. T., and Mandell, A. J. (1972): Propranolol-induced rapid activation of rat striatal tyrosine hydroxylase concomitant with behavioral depression. *Biological Psychiatry,* 4:193–203.

Tyer, P. (1972): Propranolol in alcohol addiction. *Lancet,* 2:707.

Walle, T., and Gaffney, T. E. (1972a): N-Dealkylation and oxidative deamination of propranolol in the cardiopulmonary circuit of dogs. *Abstracts of Volunteer Presentations, Fifth International Congress on Pharmacology,* p. 245. San Francisco, California.

Walle, T., and Gaffney, T. E. (1972b): Propranolol metabolism in man and dog: Mass spectrometric identification of six new metabolites. *Journal of Pharmacology and Experimental Therapeutics,* 182:83–92.

Walle, T., Ishizaki, T., and Gaffney, T. E. (1973): Isopropylamine, a biologically active deamination product of propranolol in dogs: Identification of deuterated and unlabeled isopropylamine by gas chromatography-mass spectrometry. *Journal of Pharmacology and Experimental Therapeutics, in press.*

Walle, T., Ishizaki, T., Saelens, D., Privitera, P., Garteiz, D., and Gaffney, T. E. (1972): Propranolol metabolism in man. Pharmacological properties of new metabolites. *Abstracts of Invited Presentations, Fifth International Congress on Pharmacology,* p. 74. San Francisco, California.

Wheatly, D. (1969): Comparative effects of propranolol and chlordiozepoxide in anxiety states. *British Journal of Psychiatry,* 115:1411.

Advances in Biochemical Psychopharmacology, Vol. 7
Raven Press, New York © 1973

The Use of Gas Chromatography-Mass Spectrometry Methods in Perinatal Pharmacology

M. G. Horning, D. J. Harvey, J. Nowlin,
W. G. Stillwell, and R. M. Hill

*Institute for Lipid Research and Department of Pediatrics, Baylor College of Medicine,
Houston, Texas 77025*

I. INTRODUCTION

The placental transfer of drugs and drug metabolites has been studied most extensively during labor and delivery. Much less information is available with respect to the total exposure of the human fetus to pharmacologically active compounds throughout the gestational period, and of the neonate after delivery, although with the increasing use of amniocentesis it should be possible to confirm the placental transfer of drugs ingested by the mother during most of the gestational period. After delivery, the exposure of the infant to drugs and drug metabolites may continue. Drugs are often present in breast milk, and, although the concentration of drugs may be low, the total daily ingestion may approach levels that are therapeutic for the neonate. For infants that are not breast fed, cow's milk may, in some instances, be a source of potentially toxic agents since alkaloids, drugs, pesticides, and herbicides have been identified in cow's milk. (Rasmussen, 1966).

Earlier studies of perinatal pharmacology have been limited by the low concentration of drugs and the small size of the biological sample. With the development of selective and sensitive analytical methods based on gas chromatographic-mass spectrometric (GC-MS) procedures, it is now possible to identify and quantify most drugs and drug metabolites present in amniotic fluid, breast milk, blood, and urine. The question to be answered by additional studies is not whether drugs are transferred from the mother to the fetus or neonate (they are), but rather how extensive and potentially damaging is the resultant exposure of the fetus and neonate to pharmacologically active agents.

II. METHODS

Analytical procedures have been developed for isolating drugs and drug metabolites from urine, plasma, breast milk, and amniotic fluid. These procedures have been described in detail in earlier publications from this laboratory (Harvey, Glazener, Stratton, Nowlin, Hill, and Horning, 1972*b*; Horning, Boucher, Stafford and Horning, 1972; Horning, Nowlin, Hickert, Stillwell, and Hill, 1973*a*). After isolation, using either the DEAE-Sephadex (Jaakonmaki, Knox, Horning, and Horning, 1967) or the ammonium carbonate-isopropanol extraction procedure (Horning et al., 1973*a*), the isolated drugs and metabolites are converted to derivatives (methylation followed by silylation) suitable for analysis by gas chromatography (GC) or gas chromatography-mass spectrometry (GC-MS). Established GC procedures are used for most quantitative analyses of urine. Quantitative analyses of breast milk, amniotic fluid, and plasma are carried out by selective ion detection using an LKB 9000-PDP 12 (GC-MS-COM) system operated in the electron impact mode. All of the labeled peaks in the figures have been identified by GC-MS.

A. Gas Chromatography-Mass Spectrometry

Gas chromatographic separations were carried out with 12 ft × 4 mm glass W-columns containing 5% SE-30 on 80 to 100 mesh acid washed and silanized Gas Chrom P. Mass spectrometric studies were carried out with an LKB 9000 GC-MS-COM system (PDP 12 computer). The gas chromatographic column was a 9 ft × 3.4 mm glass coil with a 1% SE-30 packing. Mass spectra were recorded at 70 eV with an accelerating voltage of 3.5 kV and an ionizing current of 60 μA. The ion source temperature was 250°C. All GC (including GC-MS) separations were carried out by temperature programming from 90°C at 2°/min.

Selective ion monitoring studies for diphenylhydantoin were carried out under isothermal conditions at 208°C. The ion m/e 180 (the base peak in the electron impact spectrum of diphenylhydantoin) was monitored. Secobarbital was monitored at m/e 196 under isothermal conditions at 120°C.

III. RESULTS AND DISCUSSION

The quantitative excretion of drugs and drug metabolites has been studied in the neonate. Most infants at birth have drugs and drug metabolites (in μg to mg quantities) stored in neonatal tissues as a result of placental transfer. These foreign compounds are excreted by the neonate in the first few days after delivery. Figure 1 shows a drug profile of urine collected from a neonate 9 to 17 hr after delivery. The mother of this infant was on chronic phenobarbital therapy and had received demerol and secobarbital during labor and delivery; caffeine had been ingested as coffee or Coca Cola. This can be considered a typical urinary drug profile for a neonate since, in our study, 98% of the mothers receive demerol and 35% receive secobarbital during labor and delivery; 98% have ingested caffeine during this period either as coffee, Coca Cola, or an analgesic preparation. A urine sample was collected from the mother at delivery, and the urinary drug profile of the mother is shown in Fig. 2. A comparison of Figs. 1 and 2 reveals the striking similarity of the maternal and neonatal drug profiles. This is not a unique example. Similar results have been obtained with many mother-infant pairs.

The time required for elimination of drugs and their metabolites from neonatal tissues depends on parameters yet to be fully defined for the newborn. Secobarbital and its metabolites acquired by placental transfer, for example, are difficult to find in neonatal urines 3 to 4 days after delivery. However, phenobarbital and its metabolites have been identified in neonatal urines 7 to 12 days after delivery.

These analyses were carried out by GC using a flame ionization detector. When more sensitive methods of detection are employed (selective ion monitoring) it will undoubtedly be found that the time required for complete elimination of drugs by the neonate is much longer.

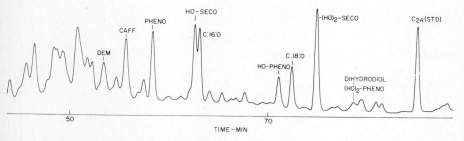

FIG.1. GC separation of the ME-TMS derivatives of a urinary fraction containing neutral and basic metabolites. The urine sample was collected from the neonate 9 to 17 hr after birth. The metabolites identified were demerol (DEM), caffeine (CAFF), phenobarbital (PHENO), hydroxysecobarbital (HO-SECO), p-hydroxyphenobarbital as the p-methoxy derivative (HO-PHENO), dihydroxysecobarbital [(HO)$_2$-SECO], the dihydrodiol of phenobarbital (DIHYDRODIOL), and dihydroxyphenobarbital [(HO)$_2$-PHENO)]. Palmitic (C:16:0) and stearic (C:18:0) acids were also present. Tetracosane (C$_{24}$STD) was added as an internal standard for quantification. The compounds were separated on a 5% SE-30 column by temperature programming at 2° min from 90°C. Demerol, caffeine, phenobarbital, and secobarbital had been ingested by the mother.

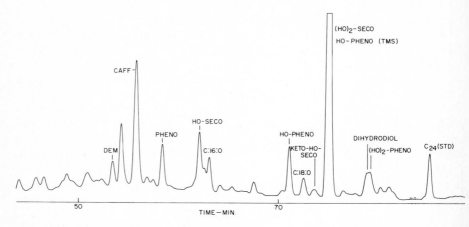

FIG. 2. GC separation of the ME-TMS derivatives of a urinary fraction containing neutral and basic metabolites. A random sample of urine (55 ml) was collected from the mother of the neonate (Fig. 1) at delivery. The separation conditions were the same as those listed in Fig. 1. Ketohydroxysecobarbital (KETO-HO-SECO) and trace amounts of the TMS ether of *p*-hydroxyphenobarbital [HO-PHENO (TMS)] as well as the drug metabolites listed in Fig. 1 were identified in the maternal urine.

The quantities of drugs and/or metabolites present in neonatal urine vary considerably. The neonatal urinary excretion may be quite high for drugs such as aspirin which are ingested by gravid females in 0.5 to 1.0 g quantities. Values in the range of 40 to 75 mg/24 hr for gentisic acid and 48 to 115 mg/24 hr of salicyluric acid have been observed in the neonate following aspirin ingestion by the mother. However, for most drugs found in neonatal urine, the amounts excreted are in the μg/24-hr range. In Table 1 the results of quantitative analyses of urine samples obtained from a neonate (Fig. 1) 9 to 17 and 17 to 25 hr after delivery are compared with the analysis of a sample of urine obtained from the mother (Fig. 2) at delivery.

All of the drugs and metabolites were excreted in μg quantities. The amount of unchanged demerol, caffeine, and phenobarbital in the neonatal urine was less in the 17 to 25 hr collection than in the 9 to 17 hr collection, and the concentration (μg/ml) had also decreased. In contrast, the total excretion and the concentration (μg/ml) of hydroxyphenobarbital [5-ethyl-5-(*p*-hydroxyphenyl)-barbituric acid], dihydroxysecobarbital [5-(2,3-dihydroxypropyl)-5-(1-methylbutyl)-barbituric acid], and ketohydroxyseco-barbital [5-(2-keto-3-hydroxypropyl)-5-(1-methylbutyl)-barbituric acid] increased in the 17 to 25 hr collection. The increase in hydroxylated metabolites is in agreement with the earlier observation that the human neonate has active enzyme systems for carrying out aliphatic and aromatic

TABLE 1. *Urinary drugs and drug metabolites*[a]

Drug or drug metabolite	9 to 17 hr (Baby D)		17 to 25 hr (Baby D)		Delivery sample (Mother D)
	μg/8 hr	μg/ml	μg/8 hr	μg/ml	μg/ml
Demerol[b]	17.4	1.9	10.7	1.4	1.3
Caffeine	29.5	3.3	18.4	2.4	7.7
Secobarbital[c]	—		—		—
Phenobarbital	29.7	3.3	16.2	2.1	2.2
Hydroxysecobarbital	35.3	3.9	27.8	3.6	3.5
Hydroxyphenobarbital[d]	13.9	1.5	33.1	4.2	2.7
Ketohydroxysecobarbital	1.7	0.2	3.6	0.5	0.5
Dihydroxysecobarbital[e]	45.5	5.1	57.8	7.4	25.1
Dihydrodiol-phenobarbital	7.2	0.8	6.6	0.8	1.9

[a] Isolated by the DEAE-Sephadex procedure. Calculations of urinary excretion were based on area relationships with respect to an internal standard (hexadecane or tetracosane) assuming a response factor of unity. The identity of all the peaks quantified by GC was confirmed by GC-MS.

[b] The recovery of demerol and caffeine is low using this procedure; the values have not been adjusted for low recovery. The recovery of barbiturates and barbiturate metabolites was 92 to 98%.

[c] The concentration of secobarbital in the isolated samples was too low to measure by GC.

[d] Measured as the methyl ether, 5-ethyl-5-(p-methoxyphenyl)-barbituric acid.

[e] This peak contained traces of hydroxyphenobarbital as the TMS ether, 5-ethyl-5-(p-trimethylsilyloxyphenyl)-barbituric acid.

hydroxylation and epoxidations from birth (Horning, Stratton, Nowlin, Wilson, Horning, and Hill, 1973*b*).

Amniotic fluid and colostrum were also obtained from this mother. The GC analysis of amniotic fluid collected 19 hr before delivery is shown in Fig. 3. Both caffeine and phenobarbital were identified by GC-MS. The concentration of phenobarbital was 2.4 μg/ml and of caffeine, 1.0 μg/ml. The GC analysis of colostrum collected 58 hr after delivery is shown in Fig. 4. The concentration of phenobarbital was 3.3 μg/ml.

Maternal and cord blood were also obtained at the time of delivery and blood was obtained from the infant 12 hr after birth. Phenobarbital was identified in all three samples. The GC analysis of neonatal plasma is shown in Fig. 5. Caffeine was present at a concentration of 0.8 μg/ml. The concentration of phenobarbital in maternal plasma at delivery was 5.3 μg/ml. The quantification of phenobarbital and caffeine was carried out by standard GC procedures. Secobarbital was identified by selective ion monitoring.

From the rather extensive series of biological samples obtained from this mother-infant pair, it was possible to show that phenobarbital, as a result of chronic administration, was present in amniotic fluid (2.4 μg/ml) and in

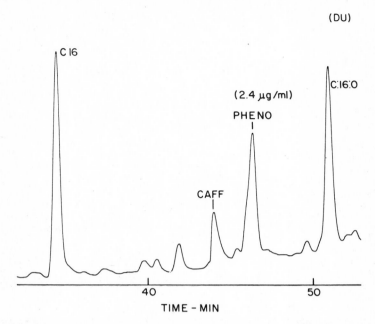

FIG. 3. GC analysis of human amniotic fluid collected 19 hr before delivery from a mother (Fig. 2) maintained on phenobarbital (195 mg/day). The compounds were separated as the methylated derivatives using conditions described in Fig. 1 with temperature programming from 100°C. The peaks identified by GC-MS were caffeine (CAFF), phenobarbital (PHENO), and palmitic acid (C:16:0).

maternal plasma (5.3 μg/ml) at delivery and in colostrum (3.3 μg/ml) 58 hr after delivery. Furthermore, the drugs and their metabolites present in the maternal urine were also present in the neonatal urine. The concentration of phenobarbital in neonatal plasma 12 hr after delivery was 5.3 μg/ml, the same as the mother's blood level at the time of delivery. Thus, the neonate was maintaining, for some time, a blood level of phenobarbital comparable to that of an adult.

Quantification of phenobarbital in all of the samples obtained from this mother-infant pair could be carried out by standard GC procedures; symmetrical GC peaks, homogeneous by GC-MS analysis, were obtained. Usually, however, it is necessary to carry out the analyses of drugs in plasma, breast milk, and amniotic fluid by selective ion monitoring. For example, diphenylhydantoin, although present in plasma and breast milk in quantities (2 to 15 μg/ml) comparable to phenobarbital, can be difficult to quantify by GC because the drug is not separated from interfering compounds found in both breast milk and plasma when the ammonium carbonate-isopropanol

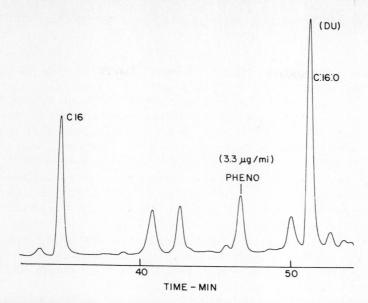

FIG. 4. GC analysis of the isopropanol extraction of human colostrum collected 58 hr after delivery from a mother (Fig. 2) maintained on phenobarbital (195 mg/day). The compounds were separated as the methylated derivatives using conditions described in Fig. 1 with temperature programming from 100°C. The peaks identified by GC-MS were phenobarbital (PHENO) and palmitic acid (C:16:0).

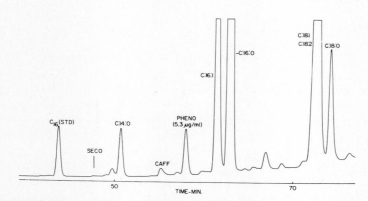

FIG. 5. GC analysis of the ethyl acetate extraction of plasma obtained from the infant (Fig. 1) 12 hr after delivery. The compounds were separated as the methylated derivatives using conditions described in Fig. 1 with temperature programming from 100°C. The peaks identified by GC-MS were caffeine (CAFF), phenobarbital (PHENO), and palmitoleic (C:16:1), palmitic (C:16:0), oleic (C:18:1), linoleic (C:18:2), and stearic (C:18:0) acids. Hexadecane (C_{16}) was added as the internal standard for the quantification of phenobarbital (5.3 μg/ml) and caffeine (0.8 μg/ml). Secobarbital (SECO) was identified by selective ion detection as described in the text.

extraction method is used. Figure 6 illustrates the problem. Further purification of the extract results in losses of diphenylhydantoin. It is possible,
however, to carry out analyses on the isopropanol extract without further
purification using the techniques of selective ion monitoring. The results of
one of the studies on breast milk are shown in Fig. 7. A series of breast milk
samples were obtained at 6:30 a.m., 1:30 p.m., 5:00 p.m., 9:45 p.m., and
11:30 p.m. from a mother maintained on diphenylhydantoin (300 mg/day
as three 100 mg doses/day). The concentration of diphenylhydantoin
measured by selective ion monitoring varied from 1.4 to 4.2 μg/ml. The
concentration was lowest in the 6:30 a.m. sample which was obtained 11.5
hr after the previous dose.

   Plasma drug levels can also be measured in the mother and neonate by
single- or multiple-ion detection. Secobarbital, for example, is usually
present in maternal and neonatal plasma in quantities too low (0.5 to 1.5
μg/ml) to be measured by GC methods when the sample size is limited
(Fig. 5). However, quantification of secobarbital in maternal, neonatal,
and cord plasma was carried out successfully by selective monitoring. The
concentration of secobarbital in maternal plasma at delivery was 1.55 μg/ml
and 76 hr after delivery had fallen to 0.42 μg/ml. The feasibility of measuring
maternal and neonatal plasma drug concentrations using small samples
(0.1 to 0.5 ml plasma) opens the way for many important pharmacokinetic
studies in the newborn.

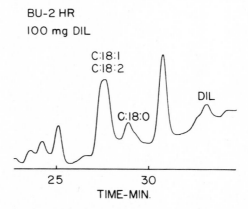

FIG. 6. GC analysis of the isopropanol extract of human breast milk collected 19 days
after delivery. The mother was maintained on 300 mg of diphenylhydantoin (DIL) per day.
The compounds were separated as the methylated derivatives using the conditions described in Fig. 1 with temperature programming from 190°C. The peaks identified by GC-
MS were C:18:1, oleic acid; C:18:2, linoleic acid; C:18:0, stearic acid; and DIL, diphenylhydantoin.

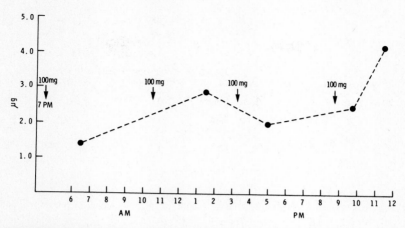

FIG. 7. Analysis of isopropanol extract of human breast milk by single-ion monitoring using an LKB-9000 mass spectrometer; the ion *m/e* 180 was monitored. Samples were obtained 19 days after delivery at 6:30 a.m., 1:30 p.m., 5:00 p.m., 9:45 p.m., and 11:30 p.m. Diphenylhydantoin (100 mg) was ingested at 10:30 a.m., 3:15 p.m. and 8:45 p.m. and at 7:00 p.m. on the previous day.

In addition to the quantitative studies that have been discussed, GC-MS methods are invaluable for the study of pathways of drug metabolism. For example, it has been reported that the newborn human and other mammals lack hepatic microsomal drug metabolizing enzymes (Fouts and Hart, 1965). However, it is possible to demonstrate by GC-MS procedures, that the human neonate does have active enzyme systems for carrying out many of the hydroxylations characteristic of the cytochrome P-450 system. When drugs are administered directly to the neonate, hydroxylated metabolites are excreted in urine. Figure 8 shows the urinary drug profile of a neonate on day 5 following administration of secobarbital and diphenylhydantoin from day 2 for medical reasons. Hydroxysecobarbital [5-allyl-5-(3-hy-droxy-1-methylbutyl)-barbituric acid], dihydroxysecobarbital [5-(2,3-dihydroxypropyl)-5-(1-methylbutyl)-barbituric acid], and hydroxydiphenyl-hydantoin [5-(*p*-hydroxyphenyl)-5-phenylhydantoin] were identified, demonstrating that active enzymes for carrying out aliphatic and aromatic hydroxylations as well as allylic epoxidation were present in the neonate.

Dihydroxysecobarbital present in neonatal urine is formed by way of an epoxide. From Figs. 1, 2, and 8, it is clear that dihydroxysecobarbital is a major metabolite in the human infant and adult. Since the epoxide of seco-barbital is converted to the diol nonenzymatically in urine, it is not possible at present to determine how much of the diol was formed *in vivo* and how much was formed nonenzymatically from the epoxide. It is of interest that

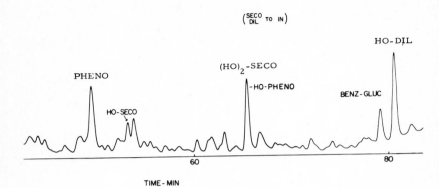

FIG. 8. GC separation of the ME-TMS derivatives of drugs and metabolites present in a solvent extract of urine. The compounds identified were phenobarbital [PHENO], hydroxyphenobarbital [HO-PHENO], hydroxysecobarbital [HO-SECO], dihydroxysecobarbital [(HO)₂-SECO], hydroxydiphenylhydantoin [HO-DIL], and benzoyl glucosiduronic acid [BENZ-GLUC]. Secobarbital and diphenylhydantoin had been given directly to the infant, whereas phenobarbital had been given to the mother. A 24-hr urine collection was made on day 5. The separation conditions were the same as those for Fig. 1.

the epoxides of secobarbital and related allylic barbiturates have been isolated as the chloro-trimethylsilyl derivatives from rat urine (Harvey, Glazener, Stratton, Johnson, Hill, Horning, and Horning, 1972a).

The dihydrodiol metabolites of aromatic drugs are probably also formed by way of an epoxide intermediate. Oesch and Daly (1971) have shown that aromatic epoxides are converted to dihydrodiols enzymatically. From GC-MS studies it is clear that many drugs containing an aromatic ring are metabolized to dihydrodiols. The dihydrodiol of diphenylhydantoin has been identified as a major unconjugated metabolite in neonatal urine (Horning, Stratton, Wilson, Horning, and Hill, 1971). The dihydrodiol of phenobarbital has been identified in the urine of rat, guinea pig, and human (Harvey et al., 1972b). The dihydrodiol of phenobarbital is present in significant quantities in the drug profile of maternal and neonatal urine shown in Figs. 1 and 2; the excretion in μg/24 hr is listed in Table 1. A new aromatic dihydrodiol has recently been identified as a metabolite of methsuximide in the rat, human, and guinea pig in studies carried out in our laboratory. The characteristic mass spectra of aromatic dihydrodiols which exhibit a base peak at m/e 191 (Horning et al., 1971) greatly facilitate their identification in GC-MS analysis of biological samples (Fig. 9). The GC-MS analytical methods have demonstrated the importance of the epoxide-diol pathway for drugs in common use.

The application of GC-MS analytical methods to studies of perinatal pharmacology have only recently been initiated. However, one can not

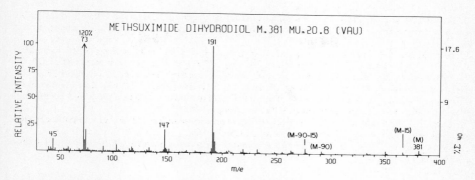

FIG. 9. Mass spectrum of the Me-TMS derivative of the dihydrodiol metabolite of meth-suximide obtained with an LKB-9000. The base peak m/e 191 has the structure $(CH_3)_3SiO$—$CH{=}\overset{+}{O}Si(CH_3)_3$; this peak was shifted by 18 amu in the perdeuterated analogue establishing the structure of the ion. A molecular ion of low intensity was present at m/e 381.

overestimate their usefulness in this area where high sensitivity of detection is essential because of the small size of the biological sample. The methods can be used for quantitative and qualitative evaluation of new and established pathways of drug metabolism and for pharmacokinetic studies of drug metabolism in the developing infant.

ACKNOWLEDGMENT

This work was supported by Grant GM-16216 of the National Institutes of General Medical Sciences.

REFERENCES

Fouts, J. R., and Hart, L. G. (1965): Hepatic drug metabolism during the perinatal period. *Annals of the New York Academy of Science*, 123:245–251.
Harvey, D. J., Glazener, L., Stratton, C., Johnson, D. B., Hill, R. M., Horning, E. C., and Horning, M. G. (1972a): Detection of epoxides of allyl-substituted barbiturates in rat urine. *Research Communications in Chemical Pathology and Pharmacology*, 4:247–260.
Harvey, D. J., Glazener, L., Stratton, C., Nowlin, J., Hill, R. M., and Horning, M. G. (1972b): Detection of a 5-(3,4-dihydroxy-1,5-cyclohexadien-1-yl)-metabolite of phenobarbital and mephobarbital in rat, guinea pig and human. *Research Communications in Chemical Pathology and Pharmacology*, 3:557–565.
Horning, M. G., Boucher, E. A., Stafford, M., and Horning, E. C. (1972): A rapid procedure for the isolation of drugs and drug metabolites from plasma. *Clinica Chimica Acta*, 37:381–386.
Horning, M. G., Nowlin, J., Hickert, P., Stillwell, W. G., and Hill, R. M. (1973a): Identification of drugs and drug metabolites in breast milk by gas chromatography-mass spectrometry. In *Dietary Lipids and Neonatal Development*, edited by C. Galli, G. Jacini, and A. Pecile. Raven Press, New York.

Horning, M. G., Stratton, C., Nowlin, J., Wilson, A., Horning, E. C., and Hill, R. M. (1973b): Placental transfer of drugs. In: *Fetal Pharmacology,* edited by L. Boreus. Raven Press, New York.

Horning, M. G., Stratton, C., Wilson, A., Horning, E. C., and Hill, R. M. (1971): Detection of 5-(3,4-dihydroxy-1,5-cyclohexadien-l-yl)-5-phenylhydantoin as a major metabolite of 5,5-diphenylhydantoin (Dilantin) in the newborn human. *Analytical Letters,* 4:537–545.

Jaakonmaki, P. I., Knox, K. L., Horning, E. C., and Horning, M. G. (1967): The characterization by gas-liquid chromatography of ethyl β-D-glucosiduronic acid as a metabolite of ethanol in rat and man. *European Journal of Pharmacology,* 1:63–70.

Oesch, F., and Daly, J. (1971): Solubilization, purification and properties of a hepatic epoxide hydrase. *Biochimica et Biophysica Acta,* 227:692–697.

Rasmussen, F. (1966): *Studies on the mammary excretion and absorption of drugs.* Copenhagen, Denmark.

Advances in Biochemical Psychopharmacology, Vol. 7
Raven Press, New York © 1973

Gas Chromatographic-Mass Spectrometric Studies on Carbamazepine

A. Frigerio, K. M. Baker, and P. L. Morselli

Istituto di Ricerche Farmacologiche Mario Negri, 20157 Milano, Italy

Mass spectrometry, particularly in combination with gas chromatography, has been shown to be particularly useful in drug metabolism and disposition studies (Hammar, Holmstedt, Lindgren, and Tham, 1969; Horning and Horning, 1971). Here we report the use of mass spectrometry for study of the fate of carbamazepine (CBZ) in both man and rat, and we also illustrate some of the ambiguous results which can be obtained using combined gas chromatography-mass spectrometry, especially with acid labile compounds.

I. THE METABOLIC DISPOSITION OF CARBAMAZEPINE

Although CBZ (I) is chemically related to imipramine, it fails to relieve the symptoms of depression. Rather it is indicated in patients with grand mal or psychomotor epilepsy (Bonduelle, Bouygues, Sallon, and Chemaly, 1964; Davis, 1964; Jongmans, 1964; Livingston, 1966; Fichsel and Heyer, 1970). Presently, CBZ is the drug of choice for the treatment of trigeminal neuralgia (Blom, 1962; Dalessio and Abbott, 1966). Until very recently little was known about the metabolic disposition of CBZ. Weist and Zicha (1967) used thin-layer chromatography to analyze the urine of patients receiving CBZ and detected seven compounds with different mobility attributable to CBZ and its metabolites. Further information on the fate of CBZ injected into animals and man is available (Morselli, Gerna, Frigerio, Zanda, and De Nadai, 1971*a;* Morselli, Gerna, and Garattini, 1971*b;* Meinardi, 1972). Other reports have shown that 10,11-dihydro-10,11-epoxy-5H-dibenz [b,f]azepine-5-carboxamide (CBZ-10.11-epoxide) (II) (Frigerio, Biandrate, Fanelli, Baker, and Morselli, 1972*a*); (Frigerio, Fanelli, Biandrate, Passerini, Morselli, and Garattini, 1972*b;* Morselli, Biandrate, Frigerio, and Garattini, 1973), iminostilbene (III) (This laboratory, *unpublished*), and 10,11-dihydro-10,11-dihydroxy-5H-dibenz[b,f]azepine-5-carboxamide, CBZ-10,11-dihydroxy (Goenechea and Hecke-Seibicke, 1972) are CBZ metabolites (Fig. 1).

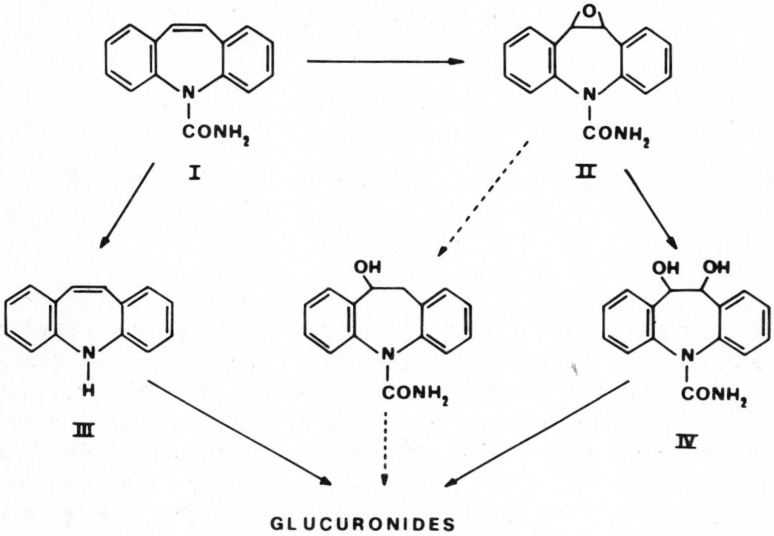

FIG. 1. Metabolic pathway for carbamazepine (I).

A. Methods

Rats housed in metabolic cages received ^{14}C-CBZ. We recovered 0.5% of the administered radioactivity in the expired CO_2 and 27% in the urine excreted in the first 24 hr. The solvent for thin-layer chromatography and the organic solvents used for urine extractions are reported in Table 1.

Since, as will be discussed later in this chapter, some of the metabolites undergo degradation during gas chromatography, the best procedure for their identification was the separation of the metabolites by thin-layer chromatography and their identification by mass spectrometry using the direct inlet. Therefore, each metabolite was characterized by its mass spectrum and by the R_f in thin-layer chromatography. When possible the identification of the metabolites was carried out in human urine because it contains larger amounts of metabolites. The gas chromatography-mass spectrometry studies reported in this chapter were performed with a magnetic deflection instrument (LKB 9000).

TABLE 1. *Thin-layer chromatography of rat and human urine
after carbamazepine (CBZ) administration*

		Rat			Man	
Standards	R_f	Extract A	Extract B	Extract C	Extract A'	Extract B'
Iminostilbene	0.70	−	+	+	−	−
CBZ	0.43	−	−	−	+	−
CBZ-10,11-epoxide	0.34	+	+	−	+	+
CBZ-10,11-dihydroxy	0.12	−	+	+	−	+

A and A' are ethylene dichloride extracts, B and B' are ethyl acetate extracts, C is ethyl acetate extract after incubation with β-glucuronidase.

R_f values of radioactive or ultraviolet, quenching zones developed in benzene-ethanol-diethylamine (8:1:1).

B. Results

The thin-layer chromatographic properties of CBZ metabolites present in the urine of man receiving 400 mg of CBZ and the rat receiving labeled CBZ are summarized in Table 1. In man and rat, we have identified the presence of CBZ-10,11-epoxide and CBZ-10,11-dihydroxy. In rat urine we also found iminostilbene. In human urine this compound was not identified but we did identify CBZ which was not present in rat urine because it had been completely metabolized.

i. Iminostilbene

Iminostilbene was identified in rat urine from its R_f and mass spectrum. The mass spectrum showed a molecular ion at m/e 193 and very little fragmentation. When urines were incubated with β-glucuronidase at pH between 4 and 5.5, the yield of iminostilbene was greatly increased. The increase did not arise from the acidic hydrolysis of CBZ because, when ^{14}C-CBZ was added to urine and was successively incubated with β-glucuronidase, we failed to detect ^{14}C-CBZ. We suggest that a large amount of iminostilbene is present in rat urine as its glucurono conjugate, and this conjugate is presumably formed during the biotransformation of CBZ.

ii. Carbamazepine (CBZ)

As shown in Table 1, the human urine contains a compound with an R_f similar to that of authentic CBZ. Moreover, the mass spectrum of this material (see Fig. 2) has a base peak at m/e 193 which corresponds to the

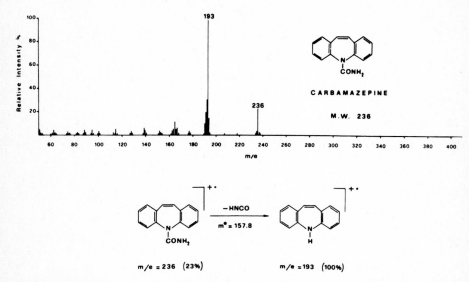

FIG. 2. Mass spectrum obtained by direct inlet system and fragmentation pathway for carbamazepine (70°C, 70 eV, 3.5 kV, LKB 9000).

molecular ion with 43 mass units accounted for by the loss of HNCO. This transition is confirmed by the presence of the corresponding metastable ion at mass 157.8.

iii. CBZ-10,11-epoxide

This metabolite accounts for 24% of the total metabolites extracted from human urine. Its identification was based on the mass spectrum identical to that of authentic material (Frigerio, et al., 1972 *b*). The mass spectrum of this urinary metabolite could be obtained only by using the direct inlet system into the ion source. As we will show later, during gas chromatography this compound is decomposed and forms 9-acridinecarboxalde-hyde (V).

As shown in Fig. 3, the mass spectrum obtained by direct injection shows a molecular ion at *m/e* 252. From this peak, with a loss of 29 mass units, a peak at *m/e* 223 was obtained. Subsequently, the latter loses 43 mass units (HNCO) to give the base peak at *m/e* 180 which corresponds to the very stable protonated acridinium ion.

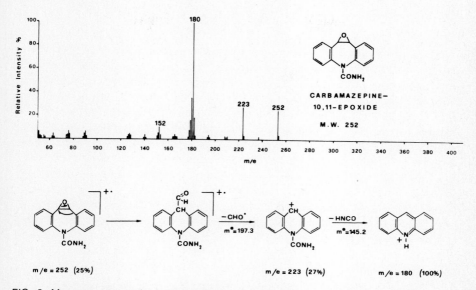

FIG. 3. Mass spectrum obtained by direct inlet system and fragmentation pathway for carbamazepine-10, 11-epoxide (70°C, 70 eV, 3.5 kV, LKB 9000).

iv. *CBZ-10,11-dihydroxy*

Recently, this metabolite has also been identified by Goenechea and Hecke-Seibicke (1972), and its structure has been confirmed mass spectrometrically in this laboratory. In urine the compound exists for the major part as the glucuronide, because the bulk of this material is extracted after β-glucuronidase incubation. CBZ-10,11-dihydroxy also undergoes degradation on a gas chromatographic column to give 9-acridinecarboxaldehyde. The structure of the dihydroxylated material can be determined from its mass spectrum which shows a molecular ion at m/e 270 with fragment ions at m/e 253 and 252 (M-OH, M-H$_2$O). The spectrum shows a base peak at m/e 180 due to the protonated acridinium ion. An authentic sample prepared by osmium tetroxide oxidation of CBZ showed the same fragmentation pattern and the same thin-layer chromatographic behavior as the material isolated from urine. The osmium tetroxide product would be expected to have a *cis* configuration for the hydroxyl groups, and, therefore, it is probable that the dihydroxylated metabolite does also.

C. DISCUSSION

CBZ and its metabolites show fragmentation patterns very characteristic of their structures. The parent compound CBZ and its metabolite iminostilbene each give intense ions at m/e 193, with little further fragmentation. The iminostilbene radical ion (VI), m/e 193, forms a very stable and completely conjugated system. This conjugation and the charge delocalization account for the extreme stability of this ion. The mass spectra of the other two metabolites, CBZ-10,11-epoxide and 10,11-dihydroxy-CBZ, show base peaks at m/e 180 which correspond to the completely conjugated protonated acridinium ion (VII). In these two cases a rearrangement of the parent ion takes place to form the conjugated system.

The stable ions (VI) and (VII) have a fragmentation pattern worthy of note. Each of them loses hydrogen cyanide, as well as the dihydrogen cyanide radical in a previously unknown concerted process (Baker and Frigerio, 1973). It is proposed that the loss of the dihydrogen cyanide radical occurs after extensive rearrangement of the parent ions as illustrated in Figs. 4 and 5 for (VI) and (VII), respectively.

The mechanisms illustrated in Figs. 4 and 5 propose the formation of an intermediate tropylium or azatropylium species of the same type as has been proposed for the loss of hydrogen cyanide from less complicated aromatic nitrogen heterocycles (Spiteller, 1966; Loader, Palmer, and Timmons, 1967; Powers, 1968; Safe, Jamieson, and Hutzinger, 1972).

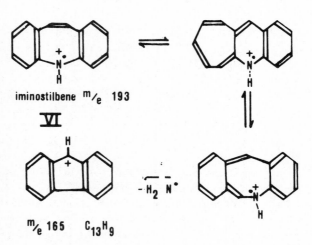

iminostilbene m/e 193

VI

m/e 165 $C_{13}H_9$

$-H_2$ N·

FIG. 4. Mechanism for the loss of the dihydrogen cyanide radical from the iminostilbene radical ion (VI).

These mechanisms resemble those proposed for quinoline (Porter and Baldas, 1971) involving substituted cyclobutadienes as product ions.

The extensive rearrangements proposed may explain why the loss of the dihydrogen cyanide radical in CBZ-10,11-epoxide and hydrogen cyanide from acridine is less abundant (9% and 7% respectively) than the loss of hydrogen cyanide from quinoline (20%) although other factors may affect the abundance of ions.

CBZ and several of its metabolites are unstable under the conditions of gas chromatography (GC) and undergo acid-catalyzed rearrangements or degradations on the column when elevated temperatures are used. When CBZ is injected in GC as a solution in methanol, it gives, in addition to a major peak due to CBZ, two peaks less polar and smaller than the CBZ peak. If ethanolic or acetonic CBZ solutions are injected, such degradations fail to occur. Gas chromatography-mass spectrometry shows that the two less polar peaks are accounted for by the formation of iminostilbene (III) and 9-methylacridine (VIII). The methanol present in the system amplifies the acidity of the column thus CBZ is hydrolyzed to form iminostilbene and, after an extensive rearrangement, 9-methylacridine.

(III)   (VIII)

The CBZ-10,11-epoxide that is formed as a metabolite of CBZ is extremely unstable to the acidity and the high temperature of the column and a complete rearrangement of the molecule occurs and 9-acridinecarboxaldehyde is formed. The structure was confirmed by the retention time and mass spectra obtained by gas chromatography-mass spectrometry (Baker, Frigerio, Morselli, and Pifferi, 1973). A rationalization of the rearrangement mechanism is shown in Fig. 6. The dihydroxylated metabolite, CBZ-10,11-dihydroxy, also undergoes a similar rearrangement under the conditions of gas chromatography. Injection of a methanol solution of CBZ-10,11-dihydroxy onto an OV 17 column gives a single-peak nonpolar material identical with 9-acridinecarboxaldehyde. The diol undergoes a pinacol-type rearrangement as shown in Fig. 6, but the rearrangement of the diol, unlike that of the CBZ-10,11-epoxide, does not take place on an

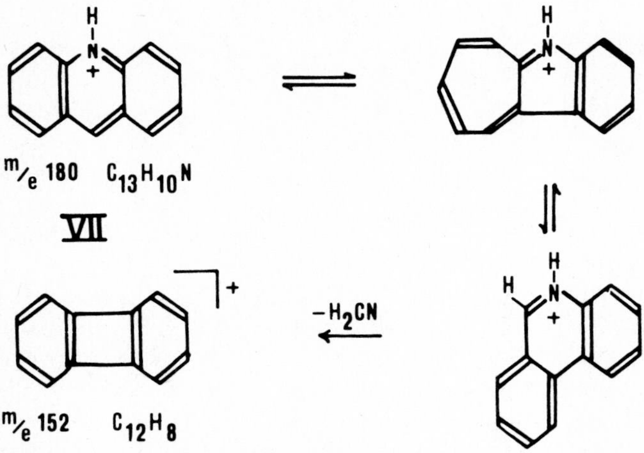

FIG. 5. Mechanism for the loss of the dihydrogen cyanide radical from the protonated acridinium ion (VII).

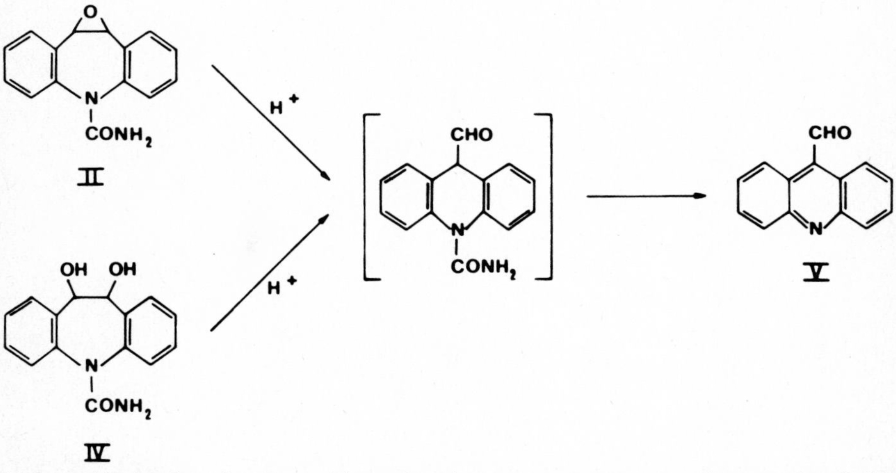

FIG. 6. Mechanism for the acid catalyzed rearrangements of carbamazepine-10,11-epoxide (II) and 10,11-dihydro-10,11-dihydroxy-5H-dibenz[b,f]azepine-5-carboxamide (IV).

SE 30 column, which is much less acidic than an OV 17 column (Kruppa, *personal communication*).

It must be emphasized that when identifying compounds or drugs by gas chromatography the structure of the material in the effluent should be checked. Standard materials should be used, and it should be borne in mind that the behavior of a compound, e.g., CBZ-10,11-dihydroxy, can change from column to column. Where possible, results should be checked by other physical methods.

ACKNOWLEDGMENTS

The authors wish to thank Professor S. Garattini for his encouragement and helpful comments. This work was supported by National Institutes of Health grant No. 1PO1 GM 18376–02 PTR.

REFERENCES

Baker, K. M., and Frigerio, A. (1973): The loss of hydrogen cyanide and the dihydrogen cyanide radical from aromatic nitrogen heterocycles on electron impact. *Journal of the Chemical Society,* Perkin II, 648–649.

Baker, K. M., Frigerio, A., Morselli, P. L., and Pifferi, G. (1973): Identification of a rearranged degradation product from carbamazepine-10,11-epoxide. *Journal of Pharmaceutical Sciences,* 62:475–476.

Blom, S. (1962): Trigeminal neuralgia: Its treatment with a new anticonvulsant drug (G-32883). *Lancet,* 1:839–840.

Bonduelle, M., Bouygues, P., Sallon, C., and Chemaly, R. (1964): Bilan de l'expérimentation clinique de l'anti-épileptique. In: *Neuropsychopharmacology,* Vol. 3, Proceedings of the Third C.I.N.P. Congress, Munich, 1962, pp. 312–316. Elsevier, Amsterdam.

Dalessio, D. J., and Abbott, K. H. (1966): A new agent in the treatment of tic douloureux: A preliminary report. *Headache,* 5:103–107.

Davis, E. (1964): Clinical evaluation of a new anti-convulsant, G.32883. *Medical Journal of Australia,* 1:150–152.

Fichsel, H., and Heyer, R. (1970): Carbamazepin in der behandlung kindlicher epilepsien. *Deutsche Medizinische Wochenschrift,* 95:2367–2370.

Frigerio, A., Biandrate, P., Fanelli, R., Baker, K. M., and Morselli, P. L. (1972a): Carbamazepine-10, 11-epoxide: A metabolite of carbamazepine isolated from human urine and identified by GLC-MS. In: *Proceedings of the International Symposium on Gas Chromatography-Mass Spectrometry,* edited by A. Frigerio, pp. 389–402. Tamburini, Milan.

Frigerio, A., Fanelli, R., Biandrate, P., Passerini, G., Morselli, P. L., and Garattini, S. (1972b): Mass spectrometric characterization of carbamazepine-10, 11-epoxide, a carbamazepine metabolite isolated from human urine. *Journal of Pharmaceutical Sciences,* 61:1144–1147.

Goenechea, S., and Hecke-Seibicke, E. (1972): Beitrag zum stoffwechsel von carbamazepine. *Zeitschrift für Klinische Chemie und Klinische Biochemie,* 10:112–113.

Hammar, C.-G., Holmstedt, B., Lindgren, J. E., and Tham, R. (1969): The combination of gas chromatography and mass spectrometry in the identification of drugs and metabolites. In: *Advances in Pharmacology and Chemotherapy,* Vol. 7, edited by S. Garattini, A. Goldin, F. Hawking, and I. J. Kopin, pp. 53–89. Academic Press, New York.

Horning, E. C., and Horning, M. G. (1971): Human metabolic profiles obtained by GC and GC/MS. *Journal of Chromatographic Science,* 9:129–140.

Jongmans, J. W. M. (1964): Report on the anti-epileptic action of Tegretol. *Epilepsia,* 5:74–82.

Livingston, S. (1966): *Drug Therapy for Epilepsy.* Charles C Thomas, Springfield.

Loader, C. E., Palmer, T. F., and Timmons, C. J. (1967): Mass spectra of some heterocyclic ring systems. In: *Some Newer Physical Methods in Structural Chemistry,* edited by R. Bonnett and J. G. Davis, pp. 80–83. United Trade Press Ltd., London.

Meinardi, H. (1972): Other antiepileptic drugs. Carbamazepine. In: *Antiepileptic Drugs,* edited by D. M. Woodbury, J. K. Penry, R. P. Schmidt, pp. 487–496. Raven Press, New York.

Morselli, P. L., Biandrate, P., Frigerio, A., and Garattini, S. (1973): Pharmacokinetics of carbamazepine in rats and humans. In: *Proceedings of the Sixth Annual Meeting European Society for Clinical Investigation,* Abstract N. 114, p. 88. *European Journal of Clinical Investigation, in press.*

Morselli, P. L., Gerna, M., Frigerio, A., Zanda, G., and De Nadai, F. (1971a): Some observations on the metabolism of carbamazepine in animals and man. In: *Fifth World Congress of Psychiatry,* Abstract N. 883. La Prensa Medica Mexicana.

Morselli, P. L., Gerna, M., and Garattini, S. (1971b): Carbamazepine plasma and tissue levels in the rat. *Biochemical Pharmacology,* 20:2043–2047.

Porter, Q. N., and Baldas, J. (1971): *Mass Spectrometry of Heterocyclic Compounds.* Wiley-Interscience, New York.

Powers, J. C. (1968): Mass spectrometry of simple indoles. *Journal of Organic Chemistry,* 33:2044–2050.

Safe, S., Jamieson, W. D., and Hutzinger, O. (1972): The ion kinetic energy and mass spectra of isomeric methyl indoles. *Organic Mass Spectrometry,* 6:33–37.

Spiteller, B. (1966): Mass spectrometry of heterocyclic compounds. In: *Advances in heterocyclic chemistry,* Vol. 7, edited by A. R. Katritzky and A. J. Boulton, pp. 301–376. Academic Press, New York.

Weist, F., and Zicha, L. (1967): Dünnscichtchromatographische untersuchungen über 5-carbamyl-5H-dibenzo [b,f] azepin in harn und liquor bei neuen indikationsgebieten. *Arzneimittel-Forschung,* 17:874–875.

Advances in Biochemical Psychopharmacology, Vol. 7
Raven Press, New York © 1973

Stable Isotope Method for the Assay of Codeine and Morphine by Gas Chromatography-Mass Spectrometry. A Feasibility Study

W. O. R. Ebbighausen, J. H. Mowat, P. Vestergaard, and N. S. Kline

Research Center, Rockland State Hospital, Orangeburg, New York 10962

I. INTRODUCTION

The advantages of the mass fragmentographic technique with stable isotopes as internal standards are high sensitivity (in the picogram range), specificity due to the focusing on specific mass fragments, and good reproducibility in quantitative work because the internal stable isotope standard makes it possible to correct for losses in the preparative phase of the analysis. The use of stable isotopes in combination with gas chromatography and mass spectrometry in pharmacology has recently been authoritatively reviewed by Knapp and Gaffney (1972).

Our interest has been in the development of stable isotope, mass fragmentographic assays in the opiate field because of the considerable social relevance of such studies at this time. Suitable stable isotope standards were found fairly easy to prepare for codeine and morphine, the first two compounds we have worked with, and fragmentation patterns for suitable derivatives were relatively favorable for the mass fragmentographic technique.

II. MATERIALS AND METHODS

A. Experimental Procedure

Codeine sulfate (Merrell Co.), 45 mg corresponding to 33.9 mg codeine base, was taken in a single oral dose by a 75-kg male volunteer. No codeine had been taken for 2 months prior to the experiment. Urine was collected for the first 10 hr after intake in 2-hr portions, then after 15, 24, 48, 72, and 96 hr. All samples were stored at 4°C.

135

Blood was collected at 1-hr intervals for 7 hr and another sample was taken at 10 hr after intake. Blood was collected in tubes containing 5 drops of sodium heparin 1,000 units/ml. After centrifugation of the blood for 15 min at 3,000 rpm, the plasma layer was separated and stored at 4°C. Blank samples of urine and blood were collected before the codeine intake.

A patient with intractable pain caused by cancer and not on morphine within the week before the experiment received a 15 mg i.m. injection of morphine hydrochloride corresponding to 13.3 mg base. The patient had an indwelling catheter because of his clinical condition, and urine was collected every 2 hr up to 10 hr and for 14 hr during the night.

B. Preparation of Stable Isotope Standards

The deuterium-labeled codeine and morphine were prepared by alkylation of norcodeine and normorphine with trideuterio methyliodide by a modification of the procedure described by Clark, Pessolano, Weijlard, and Pfister (1953). The norcodeine and normorphine used in the methylation were checked by gas chromatography and were found chromatographically pure. In particular, no trace of codeine was found in the norcodeine nor was any trace of morphine found in the normorphine.

C. Trideuterio Codeine

A mixture of norcodeine (5.5 mg), sodium carbonate (5 mg), tetrahydrofurane (0.2 ml), water (0.1 ml), and trideuterio methyliodide (2.87 mg 99.5% deuterated) was stirred in a tightly stoppered vial for 4 hr at room temperature. The mixture was transferred to a 5-ml glass-stoppered centrifuge tube, the organic solvent was evaporated with a stream of nitrogen, and approximately 1.25 ml of water was added. After adding a few milligrams of sodium carbonate, the mixture was extracted seven times with 1.25-ml portions of chloroform-isoamyl alcohol (3:1). The pooled extract was dried with anhydrous sodium sulfate.

Gas chromatography (GC) showed a single sharp peak for D_3 codeine followed by a small peak for norcodeine. Both GC peaks gave satisfactory mass spectra.

D. Trideuterio Morphine

The corresponding isotopically labeled morphine was prepared in a similar manner from normorphine and trideuterio methyl iodide, but sodium bicarbonate was used instead of sodium carbonate. The yield of the deuterated morphine was approximately 40% by gas chromatography.

E. Calculation Procedures

Since the available quantities of codeine-D_3 and morphine-D_3 were too small to permit isolation of the pure compounds, the response from accurately measured aliquots of a solution of the crude products was compared with the response obtained from known amounts of pure codeine and morphine. Since the procedures were the same for codeine and morphine, codeine is taken as an example. A preliminary comparison was made by GC and the concentration of the codeine-D_3 solution was adjusted to that of the primary codeine standard solution. The comparison was repeated using the multiple ion detector (MID) under conditions similar to those later used for the assay of codeine in urine and plasma.

Differences in response which might be due to different relative total ionizations were compensated for by adjusting the amplification settings.

Before each analysis was started, a 1/1 codeine/codeine-D_3 mixture was injected into the GC three times. A correction factor for minor differences in responses was calculated and used to correct responses of the subsequent analyses. Quantities were calculated according to peak height ratios of the unknown codeine versus the codeine-D_3 secondary standard.

The actual compound used in the above described calibration procedure was HFBA derivative of codeine.

Similar procedures were followed for the assay of morphine except that the TFA derivative was used.

Since there was some response on adjacent channels, which would require one more correction factor, optimal accuracy was obtained by keeping the ratio of unknown to its standard as near to one as possible. The effect of adjusting the ratio on the precision of the assay can be seen in Fig. 5. The ratio was unadjusted in the analyses performed on November 2 and 3. The ratio was adjusted to the optimal range (0.5 to 2.0) for the November 14 analyses. The vertical bars indicate the variation between duplicate assay.

As can be seen from the Fig. 5, we found codeine concentrations in urine ranging from approximately 40 μg/ml to 0.005 μg/ml, which represents a ratio of compound under investigation/internal standard of 8,000/1. Since published standard curves for mass fragmentography claim good accuracy only over a very narrow range of compound under investigation/internal standard of 6/1 (Bertilsson, Atkinson, Althaus, Härfast, Lindgren, and Holmstedt, 1972), it seemed inappropriate to prepare a standard curve covering a range of 8,000/1.

F. Sample Preparation

All urine samples taken up to 24 hr after intake were diluted with H_2O

in a ratio 1/10. To 100 μl of this solution the deuterated internal standard was added and the volume brought up to 1 ml. The urine was then processed as described elsewhere (Ebbighausen, Mowat, and Vestergaard, 1973).

Blood samples were handled as urine samples except that 0.5 ml plasma was used. Proteins were precipitated by addition of 6 volumes of acetone. After evaporation of the aqueous acetone, the residue was hydrolyzed (Ebbighausen et al., 1973).

G. Formation of Derivatives

Heptafluorobutyric acid (HFBA) derivatives were formed by reacting the residue with 100 μl of heptafluorobutyric acid anhydride (Pierce Chemical Co.) 40% in ether. The mixture was left to react for 30 min at room temperature.

Trifluoroacetic acid (TFA) derivatives were formed by reacting the residue with 100 μl trifluoroacetic acid anhydride (Aldrich Chemical Co.) 40% in methylene chloride for 30 min. The reaction mixture was taken to dryness under a stream of nitrogen and then dissolved in 100 μl of $CHCl_3$.

H. Gas Chromatography

Gas chromatography was performed on a 6-ft column, 2-mm internal diameter. The column was silanized before packing with 5% dimethyl dichlorosilane in toluene (Pierce Chemical Co.). It was packed with 1% OV 17 on Gas Chrom Q 100/120 mesh (Applied Science Co.). The column was preconditioned for 6 hr at 250°C without carrier gas followed by 48 hr at 280°C with a flow rate of 20 ml/min of helium.

I. Settings for the Combined Gas Chromatograph/Mass Spectrometer

Column temperature was 228°C, carrier flow 20 ml He/min, flash heater was at 250°C, separator at 260°C, and the ion source at 290°C. The multiplier voltage was 3.5 kV. The trap current was 60 μA.

J. Instrumentation

The LKB #9000 combination gas chromatograph-mass spectrometer equipped with the newly developed multiple-ion detector/peak matcher (MID/PM) accessory has been used in this investigation. It allows the recordings over time of up to four ions with masses within the range 75 to 100% of the highest mass number. Each ion can be recorded as a smooth, continuous curve. The gain for each channel can be set independently to compensate for different ion intensities. All channels also have individual controls for filter/frequency background subtraction. The latter can be used to compensate for the column bleed of the gas chromatograph.

III. RESULTS

Mass spectra of some of the derivatives used in our work are shown in Figs. 1 and 2. On the basis of these spectra, m/e values and working conditions were chosen as shown in Table 1 which demonstrate the derivative used, the optimal electron energy, the ion focused on, and the relative total intensity of the fragment.

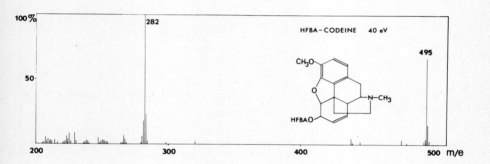

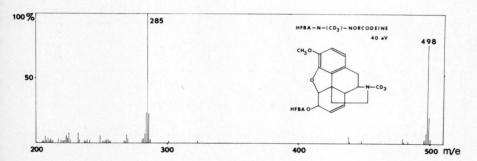

FIG. 1. Mass spectra for HFBA-codeine and HFBA-trideuterio codeine showing the high-intensity, high-mass fragments at mass numbers 495 and 498 that were chosen for mass fragmentography.

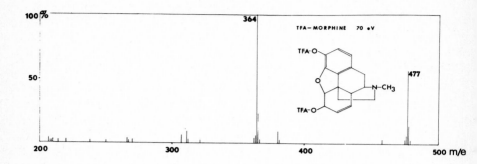

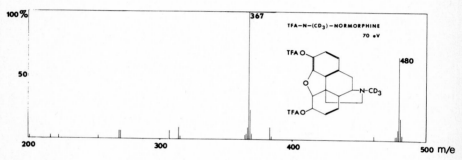

FIG. 2. Mass spectra for TFA-morphine and TFA-trideuterio morphine with the 364 and 367 fragments used in the assay.

TABLE 1. *Derivatives used in the mass fragmentographic assay and conditions of assay*

Compound	Derivative	eV	Ion focused on (rel. tot. int.)	Deuterated ion focused on
Codeine	HFBA	40	495 (M$^+$) (13.9%)	498
Morphine	TFA	70	364 (M-113) (28.0%)	367
Morphine	HFBA	70	464 (M-213) (46.4%)	467

A. Codeine in Urine and Plasma

To demonstrate the applicability of the method to biological samples, an experiment was run in which 45 mg of codeine sulfate was taken orally. Values from 0.5-ml samples of plasma gave the curve shown in Fig. 3, which is in good agreement with similar curves obtained by others using a

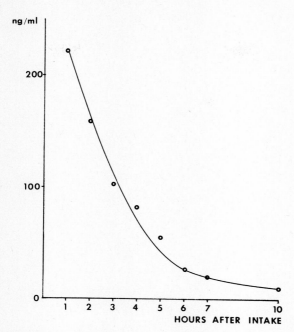

FIG. 3. Codeine concentration (in ng/ml) in plasma after a single oral dose of 45 mg of codeine sulfate.

gas chromatographic estimation of codeine in plasma (Schmerzler, Yu, Hewitt, and Greenblatt, 1966).

The curve for excretion of codeine in urine after 45-mg codeine sulfate taken orally is shown in Fig. 4. When urine samples from the first 10 hr after codeine intake were analyzed on the day after and 2 days after collection of the urine, the higher values shown on the curves to the right were obtained. To check on the stability of the codeine during storage a third estimate was made 11 days later. A fall in values is evident.

The much better agreement within duplicate estimates (indicated by the vertical bar at each value) the third time the analysis was performed is due to the optimal ratios for codeine/deuterated codeine in this set of analyses.

It can be seen from the plot that codeine was still found in the urine 72 hr after intake, although the concentration of codeine at this time was down to 5 ng/ml. The deviation from the mean of two determinations was ± 18.4%. In the 96-hr urine the codeine was still detectable, but the duplicate determinations deviated from the mean by more than 30%.

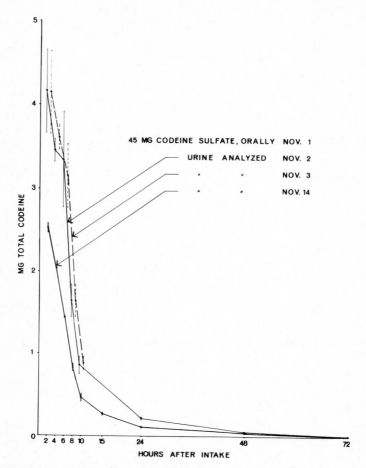

FIG. 4. Codeine excretion (in mg's) after oral intake of 45 mg of codeine sulfate. The curves from the right are from analyses done 1 and 2 days after collection. The curve to the left (analyzed 13 days after collection) had internal standard adjusted for optimal accuracy. Vertical lines around a value show difference between duplicate assays.

B. Morphine in Urine

The applicability of the stable isotope, mass fragmentographic assay to the quantitative estimation of morphine in urine was tested by taking urine samples from a patient with intractable pain after a single i.m. dose of 15 mg of morphine hydrochloride.

The single-ion curves obtained after focusing the instrument on appropriate m/e values are shown in Fig. 5. Curves are compared for a mixture of protio and trideuterio morphine, for a urine blank taken before the experiment was started, and for one of the urine samples collected after the injection. It is clear that there is minimal interference from the urine blank. The

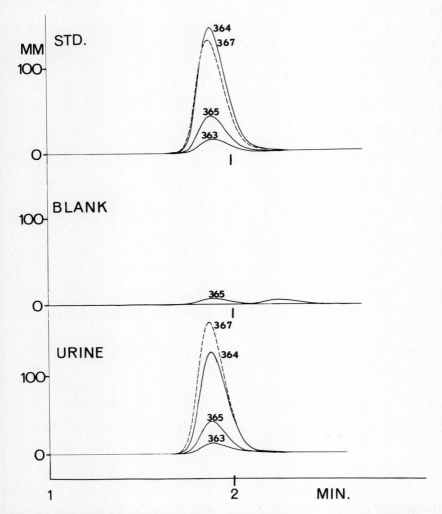

FIG. 5. Mass fragmentography of urinary morphine. Single ion recordings of mass numbers 364 and adjacent mass numbers (protio morphine) and mass number 367 (trideuterio morphine) for standard mixture (*top*), urine blank (*middle*), and urine after morphine i.m. (*bottom*).

mass numbers for morphine (364, 365, 363) are in the correct ratio in the urine when compared with the standard. Besides refocusing on m/e values, 476/477/478 confirmed the identification.

The morphine concentration in urine samples collected at different times after the injection (Fig. 6) shows a shape of the excretion curve similar to the codeine curve.

The sensitivity of this procedure for urinary morphine is similar to that found for codeine in urine and plasma, approximately 500 pg. This again is far above the sensitivity for pure compounds that is of the order of 5 to 10 pg.

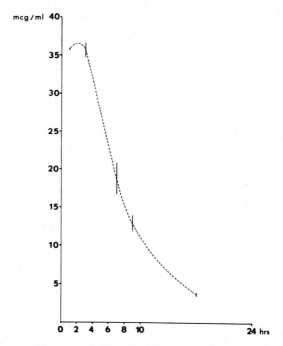

FIG. 6. Morphine concentration in urine after morphine hydrochloride 15 mg i.m. Vertical lines around a value show difference between duplicate assays.

IV. DISCUSSION

In the few months we have worked with the MID/PM accessory for the mass spectrometer we have been able to demonstrate the feasibility of assaying biological fluids for codeine and morphine. The compounds labeled with stable isotopes have been used as internal standards in mass fragmentographic assays.

Sample preparation is the main problem in developing an assay procedure for biological samples. Working with fairly crude extracts we found, for example, that a less desirable derivative from a mass fragmentographic point of view, the TFA derivative of morphine, was preferable for biological extracts because there was interference in the deuterated ion region when the generally preferred HFBA derivative was used. With the simple extraction procedure, we found the practical lower limit of detectability in the assay to be approximately 500 pg, which is several orders of magnitude higher than the 5 to 10 pg limit of detectability reached by working with standard solutions. Clearly, by further modified work-up procedures, a degree of purification may eventually be reached where the compound under analysis is essentially pure and the sensitivity found for standards will then apply.

The relative importance of sensitivity versus ease of purification will be dictated by the requirements of the problem under investigation.

For opium alkaloids, fast and simple initial extraction procedures compatible with the mass fragmentographic assay may well be found in further work. In absolute sensitivity, mass fragmentography with stable isotopes is potentially as good as any method suggested in the opiate field including the most sensitive assay yet developed in this area, the immunoassay method of Spector (1971), who claims a sensitivity for morphine of 50 to 100 pg. Besides, mass fragmentography is more specific. The maximum obtainable sensitivity in the mass fragmentographic assay seems therefore well worth striving for.

ACKNOWLEDGMENT

We want to express our appreciation to Dr. Everette L. May for supplying us with norcodeine and normorphine. This work was supported by National Institute of Mental Health grants MH 22541 and MH 20530, by institution grant GRS RR05651 from the National Institutes of Health, and by the New York State Department of Mental Hygiene.

REFERENCES

Bertilsson, L., Atkinson, A. J., Jr., Althaus, J. R., Härfast, A., Lindgren, J-E., and Holmstedt, B. (1972): Quantitative determination of 5-hydroxyindole-3-acetic acid in cerebrospinal fluid by gas chromatography-mass spectrometry. *Analytical Chemistry*, 44:1434–1438.
Clark, R. L., Pessolano, A. A., Weijlard, J., and Pfister, K. (1953): Substituted epoxymorphinans. *Journal of the American Chemical Society*, 75:4963–4967.
Ebbighausen, W. O. R., Mowat, J. H., and Vestergaard, P. (1973): Mass fragmentographic detection of normorphine in urine of man after codeine intake. *Journal of Pharmaceutical Sciences*, 62:146–148.

Knapp, D. R., and Gaffney, T. E. (1972): Use of stable isotopes in pharmacology-clinical pharmacology. *Clinical Pharmacology and Therapeutics*, 13:307–316.

Schmerzler, E., Yu, W., Hewitt, M. I., and Greenblatt, I. J. (1966): Gas chromatographic determination of codeine in serum and urine. *Journal of Pharmaceutical Sciences*, 55: 155–157.

Spector, S. (1971): Quantitative determination of morphine in serum by radioimmunoassay. *Journal of Pharmacology and Experimental Therapeutics*, 178:253–258.

Advances in Biochemical Psychopharmacology, Vol. 7
Raven Press, New York © 1973

Mass Fragmentography in Clinical Psychopharmacology

Leif Bertilsson and Folke Sjöqvist

Departments of Clinical Pharmacology (at Huddinge University Hospital), Psychiatry (at Karolinska Hospital), and Toxicology (Swedish Medical Research Council), Karolinska Institute, S-104 01 Stockholm 60, Sweden

I. INTRODUCTION

Studies of the kinetics of the tricyclic antidepressant drug nortriptyline (NT, formula in Fig. 1) showed that patients treated with fixed doses of 25 to 50 mg three times daily developed markedly different steady-state plasma concentrations (cf. Sjöqvist, Alexanderson, Åsberg, Bertilsson, Borgå, Hamberger, and Tuck, 1971). This finding raised the questions: What are the pharmacokinetic mechanisms behind these interindividual differences and what is their significance in terms of drug response? To elucidate these questions a combination of pharmacological and drug analytical techniques including gas chromatography-mass spectrometry (GC-MS) have been used. Whereas regular GC-MS requires comparatively large amounts of substance (approximately 0.5 μg), mass fragmentography (Hammar, Holmstedt, and Ryhage, 1968) has turned out to be an ideal qualitative and quantitative analytical method.

This chapter summarizes some of our applications of mass fragmentography in clinical psychopharmacology. The theory of the method is discussed elsewhere in this volume (Holmstedt and Palmér, 1973).

II. METABOLISM OF NORTRIPTYLINE IN MAN

Nortriptyline is almost completely metabolized before excretion, and interindividual differences in the biotransformation of NT were, therefore, thought to explain the variation among individuals in steady-state plasma levels of the drug.

Three metabolites of NT have been found in body fluids of man with the aid of GC-MS (Fig. 1). When low levels of the metabolites were present in plasma, mass fragmentography was used for the identification (Hammar,

147

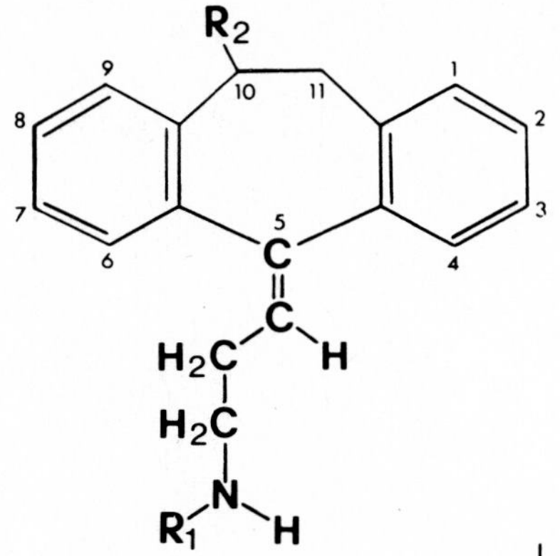

	R_1	R_2
Nortriptyline (NT)	CH_3	H
10-Hydroxynortriptyline (10-OH-NT)	CH_3	OH
Desmethylnortriptyline (DNT)	H	H
10-Hydroxydesmethyl- nortriptyline (10-OH-DNT)	H	OH

FIG. 1. Chemical structure of nortriptyline and its hitherto identified metabolites in man.

Alexanderson, Holmstedt, and Sjöqvist, 1971). Conclusive evidence was obtained that NT is hydroxylated in the 10 position and demethylated in the side chain. Tentative identification of the metabolite 10-hydroxydesmethyl-nortriptyline (10-OHDNT) was also achieved.

Figure 2 shows the log NT plasma concentrations versus time in one healthy volunteer treated with NT three times daily for 2 weeks (Alexanderson and Borgå, 1973). The excretion of conjugated and unconjugated 10-OHNT and of conjugated 10-OHDNT paralleled the disappearance of NT from plasma upon withdrawal of the drug. The concentrations of the metabolites of NT were measured with gas chromatography and electron capture detection (Borgå and Garle, 1972). The specificity of this method was assessed with GC-MS. From these studies came the conclusion that the rate of hydroxylation of NT is the major determinant for its disposition plasma half-life.

The hydroxylation of NT may result in the formation of two isomers of 10-OHNT, a *cis* and a *trans* isomer (Fig. 3). When the heptafluorobutyryl derivative of the two isomers is prepared for quantitative analysis by mass

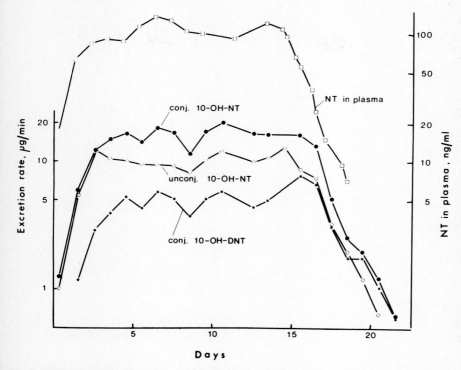

FIG. 2. Logarithmic plasma concentrations of NT versus time and log excretion rate of unconjugated and conjugated 10-OHNT and conjugated 10-OHDNT in urine during multiple-dose administration of NT   HCl orally for 2 weeks in a healthy volunteer (from Alexanderson and Borgå, 1973).

fragmentography or GC with electron capture detection, both isomers are dehydrated to the same compound $(10,11\text{-}DH_2NT)$. By these methods only total 10-OHNT can be determined. This is one of the few cases in which we have found it necessary to return' to more conventional techniques, in this case preparative thin-layer chromatography. It was then possible to separate the two isomers prior to derivative preparation (Bertilsson and Alexanderson, 1972). The heptafluorobutyryl derivative of $10,11\text{-}DH_2NT$ was then quantitated. Both isomers of 10-OHNT were formed in man. The relative proportion between the isomers in the urine was 1:4 in all six subjects studied. Differences between individuals in stereospecific hydroxylation did not explain the marked variability in steady-state plasma concentrations of NT.

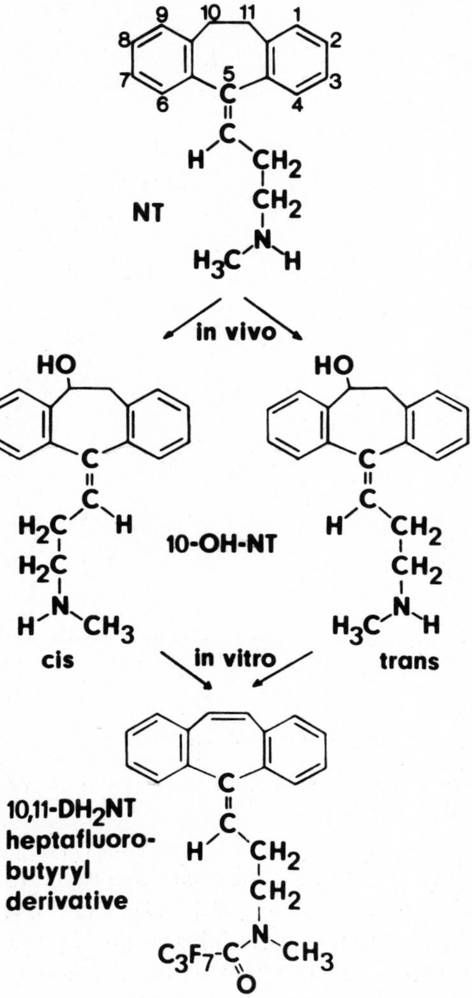

FIG. 3. Structural formulas of NT, the *cis* and *trans* isomers of 10-OHNT, and the hepta-fluorobutyryl derivative of 10,11-dehydronortriptyline (10,11-DH$_2$NT) (from Bertilsson and Alexanderson, 1972).

III. DETERMINATION OF NORTRIPTYLINE IN PLASMA

During the early part of the project NT in plasma was measured according to the principles described by Hammer and Brodie (1967) for acetylation of secondary amines with ^{3}H-acetic anhydride as further described by Sjöqvist,

Hammer, Borgå, and Azarnoff, (1969). Later a gas chromatographic method with an electron capture detector was developed (Borgå and Garle, 1972).

Concentrations of NT in plasma in the lower range (< 50 ng/ml) have also been determined with mass fragmentography (Borgå, Palmér, Linnarsson, and Holmstedt, 1971). A compound with a similar structure to NT, Ciba 34276, was used as an internal standard. To increase the accuracy of the determinations, nortriptyline labeled with three deuterium atoms was

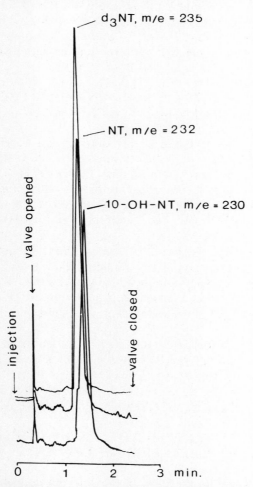

FIG. 4. Mass fragmentogram showing an analysis of NT and 10-OHNT in plasma using trideuterium-labeled NT as internal standard. The plasma contained 12 ng NT and 24 ng unconjugated 10-OHNT/ml (from Borgå et al., 1973).

later synthesized and used as an internal standard (Borgå, Palmér, Sjöqvist, and Holmstedt, 1973). Figure 4 shows a mass fragmentogram from the analysis of NT and unconjugated 10-OHNT in plasma. The mass spectrometer was set to record the base peaks of the heptafluorobutyryl derivatives of NT (m/e 232), 10-OHNT (m/e 230), and the internal standard, D_3-NT (m/e 235). By this method, NT can be determined in plasma in a concentration down to 1 ng/ml, when 2 ml of plasma is used for the analysis. These mass fragmentographic methods were developed for two main purposes: to check the specificity of the other methods for the determination of NT in plasma (Borgå et al., 1971) and to determine low concentrations of NT (< 50 ng/ml plasma) in healthy volunteers after administration of small therapeutic doses (Alexanderson and Borga, 1973).

IV. CLINICAL EFFECTS OF NORTRIPTYLINE IN RELATION TO ITS PLASMA CONCENTRATION

The relationship between the steady-state plasma level of NT and its antidepressant effect was studied in 29 inpatients with endogenous depression by Åsberg, Cronholm, Sjöqvist, and Tuck (1971). The amelioration was most pronounced in the intermediate plasma level range (50 to 139 ng NT/ml plasma) and was slight both at lower and at higher plasma levels. This curved relationship has been confirmed in another patient sample (Kragh-Sørensen, Åsberg, Eggert-Hansen, 1973). Even within the therapeutic range of plasma concentrations, some patients did not respond to this drug, which is a potent inhibitor of noradrenaline uptake but a weak inhibitor of serotonin uptake in monoaminergic neurons. There are several indications for the existence of at least two subgroups of endogenous depression, hypothetically called "noradrenaline and serotonin depressions" (cf. Sjöqvist, 1973).

We have, therefore, become interested in the question: Are there any clinical or biochemical differences between patients who respond and those who do not respond to NT at optimal pharmacokinetic conditions. To explore this hypothesis novel methods were developed for measuring metabolites of noradrenaline and serotonin in cerebrospinal fluid.

V. BIOGENIC AMINE METABOLITES IN THE CEREBROSPINAL FLUID OF DEPRESSED PATIENTS

A. Quantitative Determination of 5-Hydroxyindole-3-Acetic Acid

In 1960 Ashcroft and Sharman showed that the concentrations of the major metabolite of 5-hydroxytryptamine (5-HT), 5-hydroxyindole-3-acetic acid (5-HIAA) (Fig. 5), were lower in the cerebrospinal fluid (CSF) of depressed patients than in controls. During the last 10 years several investigations have been performed to reproduce these results. In some studies significantly less 5-HIAA was found in CSF of depressed patients than in controls but in others the results were equivocal (cf. Papeschi and McClure, 1971). In each of these investigations fluorometric techniques were relied upon.

We have developed a quantitative method for the determination of 5-HIAA in CSF by mass fragmentography (Bertilsson, Atkinson, Althaus, Härfast, Lindgren, and Holmstedt, 1972). The methyl ester of 5-HIAA is prepared with diazomethane and then acylated with heptafluorobutyryl-imidazole. This derivative contains two heptafluorobutyryl groups, one on the phenolic group and one on the indole nitrogen. The molecular ion (m/e 597) is approximately 50% of the base peak (m/e 538), which is formed by the loss of the carbomethoxy group. For the mass fragmentographic analysis of 5-HIAA these two characteristic ions were monitored. In Fig. 6 mass fragmentogram C shows material extracted from CSF. A compound with the retention time of 1 min and with ions of m/e 538 and 597 (relative intensity 2:1) was recorded. These characteristics are identical with those of reference 5-HIAA (Fig. 6A).

For the quantitative analysis of 5-HIAA in CSF the internal standard technique was used. Compounds as similar to 5-HIAA as 4-hydroxyindole-

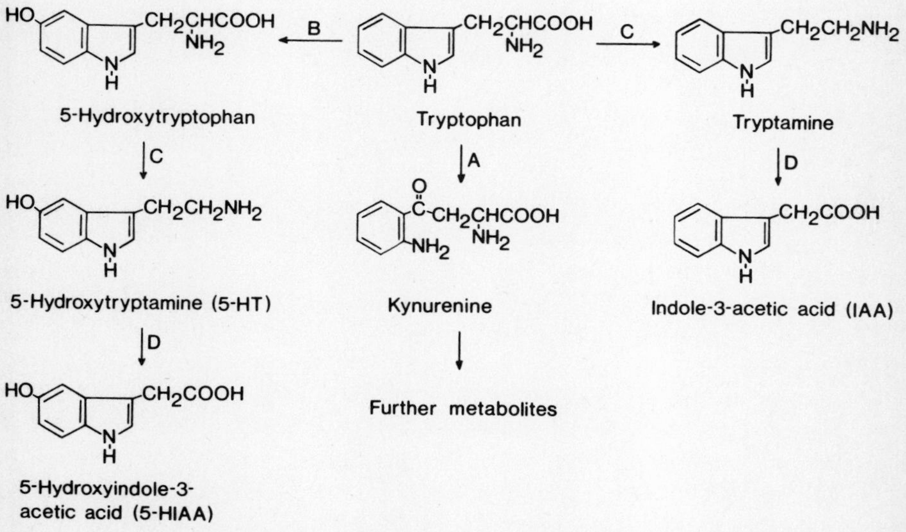

FIG. 5. Metabolism of tryptophan. The enzymes are A: tryptophan pyrrolase, B: tryptophan-5-hydroxylase, C: aromatic L-amino acid decarboxylase, D: monoamine oxidase.

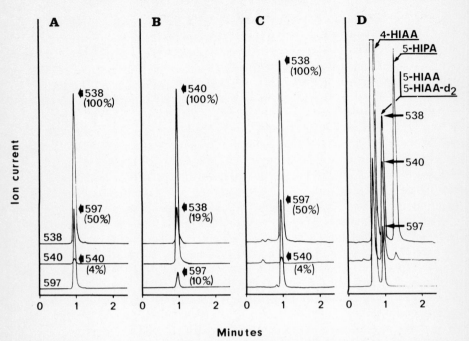

FIG. 6. Mass fragmentograms obtained from diheptafluorobutyryl methyl ester deriva-tives of A: reference 5-HIAA, B: reference 5-HIAA-D$_2$, C: material extracted from CSF, D: material extracted from CSF to which the internal standard solution containing 5-HIAA-D$_2$, 4-hydroxyindole-3-acetic acid (4-HIAA) and 5-hydroxyindole-3-propionic acid (5-HIPA) had been added. The CSF sample contained 31 ng 5-HIAA/ml (from Bertilsson et al., 1972).

3-acetic acid (4-HIAA) and 5-hydroxyindole-3-propionic acid (5-HIPA) were synthesized and tried as internal standards. The points in the standard curves that were obtained were, however, too scattered probably because neither 4-HIAA nor 5-HIPA behave exactly as the sensitive molecule 5-HIAA does in the extraction and derivatization procedures.

5-HIAA labeled with two deuterium atoms (5-HIAA-D$_2$) was subse-quently synthesized and used as internal standard. The mass spectrum of the methyl ester diheptafluorobutyryl derivative of 5-HIAA-D$_2$ had the molecular ion at m/e 599 and base peak at 540. For the registration of 5-HIAA-D$_2$ by mass fragmentography, m/e 540 was used (Fig. 6B). Al-though not useful as internal standards, 4-HIAA and 5-HIPA were added to the CSF samples (cf. Fig. 6D). The addition of 4-HIAA was found to give higher yields of the 5-HIAA and 5-HIAA-D$_2$ derivatives, probably because of its antioxidant properties. 5-HIPA was used as a marker of the 538-channel.

Standard curves for the quantitative determination of 5-HIAA in CSF (Fig. 7) were prepared by calculating the peak height ratio between 5-HIAA (m/e 538) and 5-HIAA-D$_2$ (m/e 540) and plotting it against the known concentration of 5-HIAA in standard solutions of artificial CSF. The intercept of the ordinate at 0.20 is in agreement with the ratio between m/e 538 and 540 in the mass spectrum and in the mass fragmentogram (Fig. 6B) of the 5-HIAA-D$_2$ derivative. This highly sensitive and specific method allows the quantitative determination of 5-HIAA in 2 ml of CSF with a small standard deviation.

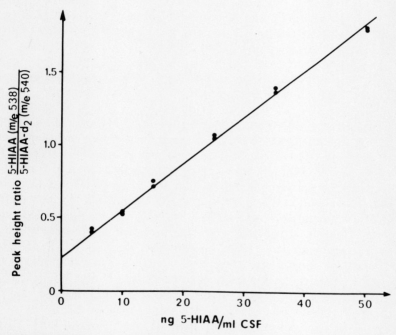

FIG. 7. Standard curve for the quantitative determination of 5-HIAA in CSF. The curve was prepared by analyzing standard solutions of 5-HIAA in artificial CSF by the entire procedure (from Bertilsson et al., 1972).

B. Identification and Quantification of Indole-3-Acetic Acid

The finding of decreased levels of 5-HIAA in CSF of depressed patients (cf. Papeschi and McClure, 1971), possibly reflecting decreased turnover of 5-HT in brain, has been hypothesized to be due to an increased tryptophan pyrrolase activity (Fig. 5) (Curzon, 1969). An increased metabolism of tryptophan through the kynurenine pathway in the periphery should leave less tryptophan available for the synthesis of 5-HT *and* tryptamine in the central nervous system. This could possibly be reflected by decreased levels of 5-HIAA and the major metabolite of tryptamine, indole-3-acetic acid (IAA), in CSF (Fig. 5).

IAA has been identified in human CSF and a quantitative method for its determination by mass fragmentography has been developed (Bertilsson and Palmér, 1972). In this method the same extraction and derivatization procedures as in the 5-HIAA analysis were used. We have used 5-methyl-indole-3-acetic acid as internal standard. The obtained standard curve (2 to 20 ng IAA/ml artificial CSF) is a straight line. 5-HIAA and IAA can be analyzed in a total volume of 2 ml of CSF. To the CSF samples, internal standards for both compounds are added.

C. Determination of 5-HIAA and IAA in CSF of Depressed Patients Before and During Nortriptyline Treatment

The methods for quantitative determination of 5-HIAA and IAA in CSF by mass fragmentography have been used in a study on 43 depressed patients (Åsberg, Bertilsson, Tuck, Cronholm, and Sjöqvist, 1973). Prior to treatment with NT a placebo was given for approximately 1 week. The patients were then given the active drug for 3 weeks (50 mg three times daily). Lumbar punctures were performed before and during treatment with NT. The mean level of 5-HIAA was 19.7 ± 7.9 ng/ml CSF before treatment. This is in agreement with earlier studies, but is somewhat lower than other results obtained with fluorometric methods (cf. Papeschi and McClure, 1971). There was a significant but weak correlation ($r = 0.46$; $p < 0.01$) between the two indoleamine metabolites in the CSF before treatment. This might be interpreted as supporting the pyrrolase theory of depression, but may be caused by the correlation of the levels of both metabolites to age. Another finding was that the levels of both 5-HIAA and IAA in CSF decreased during NT treatment. The concentrations of 5-HIAA before and during treatment were 19.2 ± 8.3 and 14.4 ± 6.8 ng/ml ($p < 0.005$), respectively. Corresponding levels of IAA were 7.5 ± 5.4 and 5.7 ± 3.4 ng/ml ($p < 0.05$). The fact that the tryptamine metabolite behaves similarly to 5-HIAA in this respect may be in keeping with the hypothesis (Dewhurst, 1968) that tryptamine is involved in central neurotransmission. The change of the 5-HIAA level in CSF during NT treatment seems ambiguous, as NT is a poor inhibitor of 5-HT uptake *in vitro* and a much more potent inhibitor of noradrenaline uptake (see below). The influence on 5-HIAA may, however, be secondary to the effect on the adrenergic neuron.

This explanation is supported by the finding that clonidine, a noradrenaline receptor stimulant, affects the turnover rate of serotonin *in vivo* (Andén, Corrodi, Fuxe, Hökfelt, Hökfelt, Rydin, and Svensson, 1970).

The distribution of the patients with endogenous depression on 5-HIAA in CSF before treatment appears to be bimodal (Fig. 8). The bipolar depressions are in both modes. This discontinuous distribution cannot be explained by differences in sex or age and we have found no clinical differences between the patients in the two modes. The tendency toward a bimodal distribution of 5-HIAA speaks in favor of a biochemical heterogeneity within the depressive syndrome, especially in the endogenous group.

The therapeutic effect of NT in those patients who had > 15 ng 5-HIAA/ml CSF (upper mode in the bimodal distribution) before treatment, was highly significant (Åsberg et al., 1973). However, the amelioration of the patients in the lower mode (< 15 ng/ml) was not significantly different from zero. All patients had apparently therapeutic plasma levels of the drug. It should be added here that accurate measurement of the very low 5-HIAA levels in CSF with fluorometry is hardly possible. This may explain why other authors have not observed a similar distribution. Nortriptyline has little effect on the neuronal uptake of 5-HT (Carlsson, Corrodi, Fuxe, and

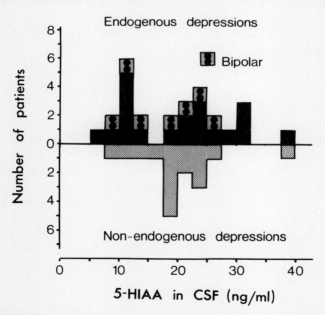

FIG. 8. Distribution of depressed patients on 5-HIAA concentration in CSF (from Åsberg et al., 1973).

Hökfelt, 1969*a,b*). Patients with low levels of 5-HIAA may benefit more from a potent inhibitor of 5-HT uptake or from treatment with tryptophan, an hypothesis which is presently under evaluation in our clinic.

To further elucidate the question about different biochemical abnormalities in the depressive syndrome, we intend to analyze both 5-HIAA and 3-methoxy-4-hydroxyphenyl glycol (MHPG) in the CSF of depressed patients.

VI. SUMMARY

This chapter does not intend to review the entire literature of mass fragmentography in clinical psychopharmacology, but rather to give examples of the application of the technique to some pertinent clinical problems in this area. The high sensitivity and selectivity of mass fragmentography has been used to quantitatively determine drugs (nortriptyline and its metabolites) and endogenous compounds (5-HIAA, IAA, and MHPG) in body fluids of man. The use of stable isotope-labeled compounds as internal standards has made the technique eminently suited for quantitative determinations in concentration ranges, where other techniques cannot be used with precision. The possibility to measure a compound and its stable isotope derivatives separately is of potential clinical interest. Thus, amino acids labeled with stable isotopes may be used in clinical studies instead of radioactive compounds. The metabolites of both the "endogenous" and "exogenous" compounds may then be analyzed.

ACKNOWLEDGMENTS

The studies discussed in this review were supported by grants from the Bank of Sweden Tercentenary Fund; the Swedish Medical Research Council (grants B73-14X-1021-08, B73-14P-3647-02, B72-14P-3589-01, B72-40Y-2375-05); the National Institutes of Health, Bethesda, Maryland (GM 13978); the National Institute of Mental Health, Chevy Chase, Maryland (grant 12007); the Wallenberg Foundation; and by funds from the Karolinska Institute.

REFERENCES

Alexanderson, B., and Borgå, O. (1973): Urinary excretion of nortriptyline and five of its metabolites in man after single and multiple oral doses. *European Journal of Clinical Pharmacology*, 5:174–180.
Andén, N. E., Corrodi, H., Fuxe, K., Hökfelt, B., Hökfelt, T., Rydin, C. and Svensson, T. (1970): Evidence for a central noradrenaline receptor stimulation by clonidine. *Life Science* 9:513–523
Asberg, M., Bertilsson, L., Tuck, D., Cronholm, B., and Sjöqvist, F. (1973): Indoleamine

metabolites in the cerebrospinal fluid of depressed patients before and during treatment with nortriptyline. *Clinical Pharmacology and Therapeutics,* 14:277–286.

Asberg, M., Cronholm, B., Sjöqvist, F., and Tuck, D. (1971): Relationship between plasma level and therapeutic effect of nortriptyline. *British Medical Journal,* 3:331–334.

Ashcroft, G. W., and Sharman, D. F. (1960): 5-Hydroxyindoles in human cerebrospinal fluid. *Nature,* 186:1050–1051.

Bertilsson, L., and Alexanderson, B. (1972): Stereospecific hydroxylation of nortriptyline in man in relation to interindividual differences in its steady-state plasma level. *European Journal of Clinical Pharmacology,* 4:201–205.

Bertilsson, L., Atkinson, Jr., A. J., Althaus, J. R., Härfast, Å., Lindgren, J.-E., and Holmstedt, B. (1972): Quantitative determination of 5-hydroxyindole-3-acetic acid in cerebrospinal fluid by gas chromatography-mass spectrometry. *Analytical Chemistry,* 44:1434–1438.

Bertilsson, L., and Palmér, L. (1972): Indole-3-acetic acid in human cerebrospinal fluid: Identification and quantification by mass fragmentography. *Science,* 177:74–76.

Borgå, O., and Garle, M. (1972): A gas chromatographic method for the quantitative determination of nortriptyline and some of its metabolites in human plasma and urine. *Journal of Chromatography,* 68:77–88.

Borgå, O., Palmér, L., Linnarsson, A., and Holmstedt, B. (1971): Quantitative determination of nortriptyline and desmethylnortriptyline in human plasma by combined gas chromatography-mass spectrometry. *Analytical Letters,* 4:837–849.

Borgå, O., Palmér, L., Sjöqvist, F., and Holmstedt, B. (1973): Mass fragmentography used in quantitative analysis of drugs and endogenous compounds in biological fluids. In: *Symposium on the Basis of Drug Therapy in Man. Fifth International Pharmacology Congress, San Francisco,* Vol. 3, C/9. S. Karger, Basel.

Carlsson, A., Corrodi, H., Fuxe, K., and Hökfelt, T. (1969*a*): Effects of antidepressant drugs on the depletion of intraneuronal brain 5-hydroxytryptamine stores caused by 4-methyl-α-ethyl-meta-tyramine. *European Journal of Pharmacology,* 5:357–366.

Carlsson, A., Corrodi, H., Fuxe, K., and Hökfelt, T. (1969*b*): Effects of some antidepressant drugs on the depletion of intraneuronal brain catecholamine stores caused by 4-α-dimethyl-meta-tyramine. *European Journal of Pharmacology,* 5:367–373.

Curzon, G. (1969): Tryptophan pyrrolase—A biochemical factor in depressive illness? *British Journal of Psychiatry,* 115:1367–1374.

Dewhurst, W. G. (1968): New theory of cerebral amine function and its clinical application. *Nature,* 218:1130–1133.

Hammar, C.-G., Alexanderson, B., Holmstedt, B., and Sjöqvist, F. (1971): Gas chromatography-mass spectrometry of nortriptyline in body fluids of man. *Clinical Pharmacology and Therapeutics,* 12:496–505.

Hammar, C.-G., Holmstedt, B., and Ryhage, R. (1968): Mass fragmentography—Identification of chlorpromazine and its metabolites in human blood by a new method. *Analytical Biochemistry,* 25:532–548.

Hammer, W., and Brodie, B. B. (1967): Application of isotope derivative technique to assay of secondary amines: Estimation of desipramine by acetylation with ^{3}H-acetic anhydride. *Journal of Pharmacology and Experimental Therapeutics,* 157:503–507.

Holmstedt, B., and Palmér, L. (1973): In: *This Volume.*

Kragh-Sørensen, P., Asberg, M., and Eggert-Hansen, C. (1973): Plasma-nortriptyline levels in endogenous depression. *Lancet,* 1:113–115.

Papeschi, R., and McClure, D. J. (1971): Homovanillic and 5-hydroxy-indoleacetic acid in cerebrospinal fluid of depressed patients. *Archives of General Psychiatry,* 25:354–358.

Sjöqvist, F. (1971): A pharmacokinetic approach to the treatment of depression. *International Pharmacopsychiatry,* 6:147–169.

Sjöqvist, F., Alexanderson, B., Asberg, M., Bertilsson, L., Borgå, O., Hamberger, B., and Tuck, D. (1971): Pharmacokinetics and biological effects of nortriptyline in man. *Acta Pharmacologia et Toxicologia,* 29: Suppl. 3, 255–280.

Sjöqvist, F., Hammer, W., Borgå, O., and Azarnoff, D. L. (1969): Pharmacological significance of the plasma level of monomethylated tricyclic antidepressants. *Excerpta Medica International Congress Series,* 180:128–136.

Advances in Biochemical Psychopharmacology, Vol. 7
Raven Press, New York © 1973

List of References of Single- and Multiple-Ion Detection

To the best of our knowledge this list contains all references to complete publications of mass fragmentography. In addition some other references related to the subject have been included. The literature search was terminated January, 1973. — THE EDITORS

Admirand, W. H., Cronholm, T., and Sjövall, J. (1970): Reduction of dehydroepiandrosterone sulfate in the liver during ethanol metabolism. *Biochimica et Biophysica Acta,* 202:343–348.

Agurell, S. (1970): The botany and chemistry of cannabis. In: *Chemical and Pharmacological Studies of Cannabis,* edited by C. R. B. Joyce and S. H. Curry. J. & A. Churchill, London.

Alexanderson, B. (1972): On interindividual variability in plasma levels of nortriptyline and desmethylimipramine in man: A pharmacokinetic and genetic study. *Linköping University Medical Dissertations,* No. 6.

Alexanderson, B. (1973): Predictions of steady-state plasma levels of nortriptyline from single oral dose kinetics, studied in twins. *European Journal of Clinical Pharmacology, in press.*

Arsenault, G. P. (1972): Chemical ionization mass spectrometry. In: *Biochemical Applications of Mass Spectrometry,* edited by G. R. Waller. Wiley-Interscience, New York.

Axen, U., Green, K., Hörlin, D., and Samuelsson, B. (1971): Mass spectrometric determination of picomole amounts of prostaglandins E_2 and $F_{2\alpha}$ using synthetic deuterium labeled carriers. *Biochemical and Biophysical Research Communications,* 45:519–525.

von Bahr, C. (1972): Binding and oxidation of amitriptyline and a series of its oxidized metabolites in liver microsomes from untreated and phenobarbital-treated rats. *Xenobiotica,* 2:293–306.

Baillie, T. A., Brooks, C. J. W., and Middleditch, C. S. (1972): Comparison of corticosteroid derivates by gas chromatography-mass spectrometry. *Analytical Chemistry,* 44:30–37.

Bergstedt, L., and Widmark, G. (1970): Repetitive scanning in gas chromatography — mass spectrometry. *Chromatographia,* 3:59–63.

Bertilsson, L., Atkinson, Jr., A. J., Althaus, J. R., Härfast, Å., Lindgren, J.-E., and Holmstedt, B. (1972): Quantitative determination of 5-hydroxyindole-3-acetic acid in cerebrospinal fluid by gas chromatography — mass spectrometry. *Analytical Chemistry,* 44:1434–1438.

Bertilsson, L., and Palmér, L. (1972): Indole-3-acetic acid in human cerebrospinal fluid: Identification and quantification by mass fragmentography. *Science,* 177:74–76.

Bieber, M. A., Sweeley, C. C., Faulkner, D. J., and Petersen, M. R. (1972): Purification and quantitative determination of *Cecropia* juvenile hormone. *Analytical Biochemistry,* 47:264–272.

Bonelli, E. J. (1972): Gas chromatograph/mass spectrometer techniques for determination of interferences in pesticide analysis. *Analytical Chemistry,* 44:603–606.

Bonelli, E. J., Story, M. S., and Knight, J. B. (1971): Computerised gas chromatography and quadrupole mass spectrometry. *Dynamic Mass Spectrometry,* 2:177–202.

Bonnichsen, R., Maehly, A. C., Marde, Y., Ryhage, R., and Schubert, B. (1970): Identification of small amounts of barbiturate sedatives in biological samples by a combination of gas chromatography and mass spectrometry. *Zacchia; Archivio di medicina legale, sociale e criminologica,* VI:3.

Borgå, O., Palmér, L., Linnarson, A., and Holmstedt, B. (1971): Quantitative determination of nortriptyline and desmethylnortriptyline in human plasma by combined gas chromatography-mass spectrometry. *Analytical Letters*, 4:837–849.

Borgå, O., Palmér, L., Sjöqvist, F., and Holmstedt, B. (1973): Mass fragmentography used in quantitative analysis of drugs and endogenous compounds in biological fluids. In: *Symposium on the Basis of Drug Therapy in Man*. Fifth International Pharmacology Congress, San Francisco, Vol. 3, C/9. S. Karger, Basel, *in press*.

Boulton, A. A., and Majer, J. R. (1970): Mass spectrometry of crude biological extracts. Absolute quantitative detection of metabolites at the submicrogram level. *Journal of Chromatography*, 48:322–327.

Boulton, A. A., and Majer, J. R. (1971): The mass spectrometric identification of the isomers of tyramine: Identification of paratyramine in rat brain. *Canadian Journal of Biochemistry*, 49:993–998.

Brandenberger, H. (1970): Die Gaschromatographie in der toxikologischen Analytik. Verbesserung von Selektivität und Empfindlichkeit durch Einsatz multipler Detektoren. *Pharmaceutica Acta Helvetiae*, 45:394–413.

Brandenberger, H. (1972): The search for drugs of abuse by gas chromatography with mass specific detection systems. In: *Proceedings of the International Symposium on Gas Chromatography Mass Spectrometry*, edited by A. Frigerio. Tamburini Editore, Milano.

Brandenberger, H., and Schnyder, D. (1972): Toxikologischer Spurennachweis von biologisch wirksamen Aminen (Amphetamine, Catecholamine, Halluzinogene) durch Gas-Chromatographie mit massenspezifischer Detektion (GC-MD). *Zeitschrift für Analytische Chemie*, 259:210–211.

Breimer, D. D., Vree, T. B., Van Ginneken, C. A. M., Henderson, P. T., and Van Rossum, J. M. (1972): Identification of cannabis metabolites by a new method of combined gas chromatography-mass spectrometry. In: *Proceedings of the International Symposium on Gas Chromatography Mass Spectrometry*, edited by A. Frigerio. Tamburini Editore, Milano.

von Breuer, H., Nocke. L., and Siekmann, L. (1970): Neue Methoden zur Bestimmung von Steroidhormonen. *Zeitschrift für Klinische Chemie und Klinische Biochemie*, 8:329–338.

Brooks, C. J. W., and Middleditch, B. S. (1971): The mass spectrometer as a gas chromatographic detector. *Clinica Chimica Acta*, 34:145–157.

Bush, M. T., Sekerke, Jr., H. J., Vore, M., Seetman, B. J., and Watson, J. T. (1971): Labeling with ^{15}N or ^{13}C for identification of mass spectra of drugs and their metabolites. In: *Proceedings of a Seminar on the Use of Stable Isotopes in Clinical Pharmacology*, edited by P. D. Klein and L. J. Roth. Argonne National Laboratory, National Technical Information Service, U.S. Department of Commerce, 5285 Port Royal Road, Springfield, Virginia 22151, Document number CONF-711115.

Caprioli, R. M. (1972): Use of stable isotopes. In.: *Biochemical Applications of Mass Spectrometry*, edited by G. R. Waller. Wiley-Interscience, New York.

Cattabeni, F., Koslow, S. H., and Costa, E. (1972*a*): Gas Chromatography-mass fragmentography: A new approach to the estimation of amines and amine turnover. In: *Advances in Biochemical Psychopharmacology*, Vol. 6, edited by E. Costa, L. L. Iversen, and R. Pauletti, Raven Press, New York.

Cattabeni, F., Koslow, S. H., and Costa, E. (1972*b*): Gas chromatographic-mass spectrometric assay of four indole alkylamines of rat pineal. *Science*, 178:166–168.

Cattabeni, F., Revuelta, A., and Costa, E. (1972*c*): Effects of β-methoxy derivatives of 3-trifluoromethylphenylethylamine on food intake and brain serotonin content. *Neuropharmacology*, 11:758–760.

Chapman, J. R., Compson, K. R., Done, D., Merren, T. O., and Tennant, P. W. (1972): New techniques in single and multiple ion monitoring used in quantitative steroid estimation. In: *Twentieth Annual Conference on Mass Spectrometry and Allied Topics*, pp. 166–171. American Society for Mass Spectrometry, Dallas.

Cho, A. K., Lindeke, B. and Hodson, B. J. (1972*a*): The N-hydroxylation of phentermine (2-methyl-1-phenylisopropylamine) by rabbit liver microsomes. *Research Communications in Chemical Pathology and Pharmacology*, 4:519–528.

Cho, A. K., Lindeke, B., Hodson, B. J., and Jenden, D. J. (1972b): A gas chromatography/mass spectrometry assay for amphetamine in plasma. In: *Proceedings of the Fifth International Congress on Pharmacology*, p. 41. S. Karger, Basel.

Cho, A. K., Lindeke, B., Hodson, B. J., and Jenden, D. J. (1973): Deuterium substituted amphetamine as an internal standard in a gas chromatographic/mass spectrometric (GC/MS) assay for amphetamine. *Analytical Chemistry*, 45:570–574.

Costa, E., Green, A. R., and Koslow, S. H. (1972a): Mass fragmentographic quantitation and multiple ion detection of endogenous catecholamines. In: *Proceedings of the International Symposium on Gas Chromatography Mass Spectrometry*, edited by A. Frigerio. Tamburini Editore, Milano.

Costa, E., Green, A. R., Koslow, S. H., LeFevre, J. F., Revuelta, A. V., and Wang, C. (1972b): Dopamine and norepinephrine in noradrenergic axons: A study *in vivo* of their precursor product relationship by mass fragmentography and radiochemistry. *Pharmacological Reviews*, 24:167–190.

Costello, C. E., Sakai, T., and Biemann, K. (1972): Identification of drugs in body fluids, particularly in emergency cases of acute poisoning. In: *Twentieth Annual Conference on Mass Spectrometry and Allied Topics*, pp. 107–111. American Society for Mass Spectrometry, Dallas.

Danzinger, R. G., Hofmann, A. F., and Schoenfield, L. J. (1971): Measurement of bile acid kinetics in man using stable isotopes: Application to cholelithiasis. *Gastroenterology*, 60:192.

DoAmaral, J. R., and Barchas, J. D. (1972): Approach to assay of biogenic amines using gas-liquid as a model system. In: *Proceedings of the Fifth International Congress on Pharmacology*, p. 58. S. Karger, Basel.

Draffan, G. H., Clare, R. A., and Williams, F. M. (1973): Determination of barbiturates and their metabolites in small plasma samples by gas chromatography-mass spectrometry. *Journal of Chromatography*, 75:45–53.

Ebbighausen, W. O. R., Mowat, J. H., and Vestergaard, P. (1973): Mass fragmentographic detection of normorphine in urine of man after codeine intake. *Journal of Pharmaceutical Sciences*, 62:146–148

Elkin, K., Pierrou, L., Ahlborg, U. G., Holmstedt, B., and Lindgren, J.-E. (1973): Computer controlled mass fragmentography with digital signal processing. *Journal of Chromatography, in press.*

Frew, N. M., and Isenhour, T. L. (1972): Mass spectrometric isotope ratio measurements and peak area integration using the peak-switching feature of the AEI MS-902. *Analytical Chemistry*, 44:659–664.

Frigerio, A., Belvedera, G., De Nadai, F., Fanelli, R., Pantarotto, C., Riva, E., and Morselli, P. L. (1972a). A method for the determination of imipramine in human plasma by gas-liquid chromatography-mass fragmentography. *Journal of Chromatography*, 74:201–208.

Frigerio, A., Belvedera, G., De Nadai, F., Fanelli, R., Pantarotto, C., Riva, E., and Morselli, P. L. (1972b): Quantitative determination of imipramine in human plasma by gas-liquid chromatography-mass fragmentography. In: *Proceedings of the International Symposium on Gas Chromatography Mass Spectrometry*, edited by A. Frigerio. Tamburini Editore, Milano.

Gaffney, T. E., Hammar, C.-G., Holmstedt, B., and McMahon, R. E. (1971): Ion specific detection of internal standards labelled with stable isotopes. *Analytical Chemistry*, 43:307–310.

Van Ginneken, C. A. M., Vree, T. B., Breimer, D. D., Thijssen, H. W. H., and Van Rossum, J. M. (1972): Cannabinodiol, a new hashish constituent, identified by gas chromatography-mass spectrometry. In: *Proceedings of the International Symposium on Gas Chromatography Mass Spectrometry*, edited by A. Frigerio. Tamburini Editore, Milano.

Glover, D. D. (1972): A multi-instrument data system for use with a GC-MS. In: *Proceedings of the International Symposium on Gas Chromatography Mass Spectrometry*, edited by A. Frigerio. Tamburini Editore, Milano.

Gohlke, R. S. (1962): Time-of-flight mass spectrometry: Application to capillary column gas chromatography. *Analytical Chemistry*, 34:1332–1333.

Gordon, A. E., and Frigerio, A. (1972): Mass fragmentography as an application of gas-liquid chromatography-mass spectrometry in biological research. *Journal of Chromatography*, 73:401–417.

Green, D. E., and Forrent, I. S. (1972): Fully automated detection and assay of drugs by "direct multiple-mass fragmentography," a new ultra-sensitive analytical technique. In: *Proceedings of the Fifth International Congress on Pharmacology*, p. 88. S. Karger, Basel.

Green, A. R., Koslow, S. H., Spano, P. F., and Costa, E. (1972): Identification of melatonin (M) and 5-methoxytryptamine (5-MT) in rat hypothalamus by gas chromatography-mass spectrometry (GC-MS). In: *Proceedings of the Fifth International Congress on Pharmacology*, p. 87. S. Karger, Basel.

Grostic, M. F. (1972): Drug metabolism. In: *Biochemical Applications of Mass Spectrometry*, edited by G. R. Waller. Wiley-Interscience, New York.

Gustafsson, J.-A., and Sjövall, J. (1969): Steroid analysis by gas chromatography-mass spectrometry. In: *Progress in Endocrinology, Proceedings of the Third International Congress of Endocrinology, Mexico, D.F.*, edited by C. Gual, pp. 470–475. ICS 184, Excerpta Medica, Amsterdam.

Hammar, C.-G. (1971): Mass fragmentography and elemental analysis by means of a new and combined multiple ion detector-peak matcher device. *Acta Pharmaceutica Suecica*, 8:129–152.

Hammar, C.-G. (1972): Qualitative and quantitative analyses of drugs in body fluids by means of mass fragmentography and a novel peak matching technique. In: *Proceedings of the International Symposium on Gas Chromatography Mass Spectrometry*, edited by A. Frigerio, pp. 1–18. Tamburini Editore, Milano.

Hammar, C.-G., Alexanderson, B., Holmstedt, B., and Sjöqvist, F. (1971): Gas chromatography-mass spectrometry of nortriptyline in body fluids of man. *Clinical Pharmacology and Therapeutics*, 12:496–505.

Hammar, C.-G., Hanin, I., Holmstedt, B., Kitz, R. J., Jenden, D. J., and Karlén, B. (1968a): Identification of acetylcholine in fresh rat brain by combined gas chromatography-mass spectrometry. *Nature*, 220:915.

Hammar, C.-G., and Hessling, R. (1971): Novel peak matching technique by means of a new and combined multiple ion detector-peak matcher device. *Analytical Chemistry*, 43:298–306.

Hammar, C.-G., and Holmstedt, B. (1967): Gaskromatografi och masspektrometri vid bestämning av psykofarmaka i kroppsvätskor. In: *Symposium om depressions-behandling*, edited by B. Cronholm and F. Sjöqvist. Lidingö, Sweden.

Hammar, C.-G., Holmstedt, B., Lindgren, J.-E., and Tham, R. (1969): The combination of gas chromatography and mass spectrometry in the identification of drugs and metabolites. *Advances in Pharmacology and Chemotherapy*, 7:53–89.

Hammar, C.-G., Holmstedt, B., and Ryhage, R. (1968b): Mass fragmentography. Identification of chlorpromazine and its metabolites in human blood by a new method. *Analytical Biochemistry*, 25:532–548.

Hanin, I., Massarelli, R., and Costa, E. (1972): An approach to the *in vivo* study of acetylcholine turnover in rat salivary glands by radio gas chromatography. *Journal of Pharmacology and Experimental Therapeutics*, 181:10–18.

Henneberg, D. (1959): Ein kontinuierliches Verfahren zur massenspektrometrischen Bestimmung gaschromatographisch vorgetrennter Substanzgemische. *Zeitschrift für Analytische Chemie*, 170:365–366.

Henneberg, D. (1961): Eine Kombination von Gaschromatograph und Massenspektrometer zur Analyse organischer Stoffgemische. *Zeitschrift für Analytische Chemie*, 183:12–23.

Henneberg, D., and Schomburg, G. (1962): Mass spectrometric identification in capillary gas chromatography. *Gas Chromatography*, 1962:191–203.

Henneberg, D., and Schomburg, G. (1966): Ein kleines Massen-spektrometer als spezifischer gaschromatographischer Detektor. *Zeitschrift für Analytische Chemie*, 215:424–430.

Hites, R. A., and Biemann, K. (1970): Computer evaluation of continuously scanned mass spectra of gas chromatographic effluents. *Analytical Chemistry*, 42:855–860.

Holmstedt, B., and Lindgren, J.-E. (1972): Nachweis von Suchtmitteln und Halluzinogenen mit Hilfe der Kombination Gas-Chromatographie/Massenspektrometrie. *Zeitschrift für Analytische Chemie*, 261:291–297.

Holmstedt, B., and Linnarson, A. (1972): Chemistry and means of determination of hallucinogens and marihuana. In: *Drug Abuse Proceedings of the International Conference,*
edited by C. J. D. Zarafonetis, pp. 291–305. Lea & Febiger, Philadelphia.

Horning, E. C., and Horning, M. G. (1972): Chemical ionization mass spectrometry in drug
metabolism studies. In: *Proceedings of the International Symposium on Gas Chromatography Mass Spectrometry,* edited by A. Frigerio. Tamburini Editore, Milano.

Jenden, D. J. (1972): Simultaneous microestimation of choline and acetylcholine by gas chro-
matography/mass spectrometry/isotope dilution. *Federation Proceedings,* 31:515.

Jenden, D. J., Booth, R. A., and Roch, M. (1972): Simultaneous microestimation of choline
and acetylcholine by gas chromatography. *Analytical Chemistry,* 44:1879–1881.

Jenden, D. J., Booth, R. A., and Roch, M. (1973): Simultaneous measurement of endogenous
and deuterium-labelled tracer variants of choline and acetylcholine in sub-picomole quantities
by gas chromatography/mass spectrometry. *Analytical Biochemistry, in press.*

Jenden, D. J., and Cambell, L. B. (1971): Measurement of choline esters. In: *Analysis of
Biogenic Amines and Their Related Enzymes,* edited by D. Glick. Interscience Publications,
New York.

Jenden, D. J., and Cho, A. K. (1972): Applications of integrated gas chromatography/mass
spectrometry in pharmacology and toxicology. *Annual Review of Pharmacology,* 13:371–
390.

Jenden, D. J., and Silverman, R. W. (1971): An analog multiplexer for multiple specific ion
detection in a gas chromatograph/quadrupole mass spectrometer system. In: *Proceedings of
a Seminar on the Use of Stable Isotopes in Clinical Pharmacology,* edited by P. D. Klein and
L. J. Roth. Argonne National Laboratory, National Technical Information Service, U.S. Department of Commerce, 5285 Port Royal Road, Springfield, Virginia 22151, Document number CONF-711115.

Jenden, D. J., and Silverman, R. W. (1973): A multiple specific ion detector and analog data
processor for a gas chromatograph/quadrupole mass spectrometer system. *Journal of
Chromatography Science, in press.*

Johansson, G., Ryhage, R., and Westöö, G. (1970): Identification and determination of methyl-
mercury compounds in fish using combination gas chromatograph-mass spectrometer. *Acta
Chemica Scandinavica,* 24:2349–2354.

Johnston, G. A. R., Triffett, A. C. K., and Wunderlich, J. A. (1968): Identification and estimation of choline derivatives by mass spectrometry. *Analytical Chemistry,* 40:1837–1840.

Karlén, B., Lundgren, G., Nordgren, I., and Holmstedt, B. (1973): Ion pair extraction in
combination with gas phase analysis of acetylcholine. In: *Advances in Neuropsychopharmacology,* edited by Z. Votava. Avicenum, Prague, *in press.*

Kelly, R. W. (1971): The measurement by gas chromatography-mass spectrometry of oestra-
1,3,5,-triene-3,15α,16α,17β-tetrol (oestetrol) in pregnancy urine. *Journal of Chromatography,* 54:345–355.

Kelly, R. W. (1972a): A simple device for analogue recording of the abundance of selected
ions from a combined gas chromatograph-mass spectrometer. *Journal of Chromatography,*
71:337–339.

Kelly, R. W. (1972b): Quantitative measurement of prostaglandins by gas chromatography-
mass spectrometry. In: *Proceedings of the International Symposium on Gas Chromatography Mass Spectrometry,* edited by A. Frigerio. Tamburini Editore, Milano.

Klein, P. D. (1971): The measurement of isotope ratios in organic molecules by mass spectroscopy. In: *Proceedings of a Seminar on the Use of Stable Isotopes in Clinical Pharmacology,* edited by P. D. Klein and L. J. Roth. Argonne National Laboratory, National
Technical Information Service, U.S. Department of Commerce, 5285 Port Royal Road,
Springfield, Virginia 22151, Document number CONF-711115.

Klein, P. D., Haumann, J. R., and Eisler, W. J. (1971): Instrument design considerations and
clinical applications of stable isotope analysis. *Clinical Chemistry,* 17:735–738.

Klein, P. D., Haumann, J. R., and Eisler, W. J. (1972): Gas chromatograph-mass spectrometer-
accelerating voltage alternator system for the measurement of stable isotope ratios in organic
molecules. *Analytical Chemistry,* 44:490–493.

Knapp, D. R., and Gaffney, T. E. (1971): Use of stable isotopes in pharmacology-clinical pharmacology. In: *Proceedings of a Seminar on the Use of Stable Isotopes in Clinical Pharmacology,* edited by P. D. Klein and L. J. Roth. Argonne National Laboratory, National Technical Information Service, U.S. Department of Commerce, 5285 Port Royal Road, Springfield, Virginia 22151, Document number CONF-711115.

Knapp, D. R., and Gaffney, T. E. (1972): Commentary. Use of stable isotopes in pharmacology-clinical pharmacology. *Clinical Pharmacology and Therapeutics,* 13:307–316.

Knapp, D. R., Gaffney, T. E., and McMahon, T. E. (1972a): Use of stable isotope mixtures as a labeling technique in drug metabolism studies. *Biochemical Pharmacology,* 21:425–429.

Knapp, D. R., Gaffney, T. E., McMahon, R. E., and Kiplinger, G. (1972b): Studies of human urinary and biliary metabolites of nortriptyline with stable isotope labeling. *Journal of Pharmacology and Experimental Therapeutics,* 180:784–790.

Kolor, M. G. (1972): Flavor components. In: *Biochemical Application of Mass Spectrometry,* edited by G. R. Waller. Wiley-Interscience, New York.

Koslow, S. H., Cattabeni, F., and Costa, E. (1972a): Norepinephrine and dopamine: Assay by mass fragmentography in the picomole range. *Science,* 176:177–180.

Koslow, S. H., Cattabeni. F., and Costa. F. (1972b): Quantitative mass fragmentography of some indolealkylamines of the rat pineal glands. In: *Pineal Gland: Proceedings of a Workshop held by the National Institutes of Health.*

Koslow, S. H., Green, A. R., and Costa, E. (1972c): Mass fragmentography quantitation and multiple ion detection of endogenous indole alkylamines. In: *Proceedings of the International Symposium on Gas Chromatography Mass Spectrometry,* edited by A. Frigerio. Tamburini Editore, Milano.

Lindeman, L. P., and Annis, J. L. (1960): Use of a conventional mass spectrometer as a detector for gas chromatography. *Analytical Chemistry,* 32:1742–1749.

Lindgren, J.-E., Agurell, S., Lundström, J., and Svensson, U. (1971): Detection of biochemical intermediates by mass fragmentography: Mescaline and tetrahydroisoquinoline precursors. *Febs Letters,* 13:21–27.

Majer, J. R., and Boulton, A. A. (1970): Absolute, unambiguous ultramicro analysis of metabolites present in complex biological extracts. *Nature,* 225:658–660.

Mikes, F., Hofmann, A., and Waser, P. G. (1971): Identification of 6a,10a-trans-tetrahydrocannabinol and two of its metabolites in rats by use of combination gas chromatography-mass spectrometry and mass fragmentography. *Biochemical Pharmacology,* 20:2469–2476.

Morgan, C. D., Cattabeni, F., and Costa, E. (1972): Methamphetamine, fenfluramine and their N-dealkylated metabolites: Effect on monoamine concentrations in rat tissues. *Journal of Pharmacology and Experimental Therapeutics,* 180:127–135.

Niehaus, Jr., W. G., and Ryhage, R. (1968): Determination of double bond positions in polyunsaturated fatty acids by combination gas chromatography-mass spectrometry. *Analytical Chemistry,* 40:1840–1841.

Palmér, L., and Kolmodin-Hedman, B. (1972): Improved quantitative gas chromatographic method for analysis of small quantities of chlorinated hydrocarbon pesticides in human plasma. *Journal of Chromatography,* 74:21–30.

Pilar, G., Jenden, D. J., and Campbell, L. B. (1973): Distribution of acetylcholine in the normal and denervated pigeon ciliary ganglion. *Brain Research, in press.*

Ratner, S. (1972): A historical survey of mass spectrometry. In: *Biochemical Applications of Mass Spectrometry,* edited by G. R. Waller. Wiley-Interscience, New York.

Reimendal, R., and Sjövall, J. (1971): Use of computer in gas chromatographic-mass spectrometric analyses of steroids. In: *Hormonal Steroids, Proceedings of the Third International Congress, Hamburg,* pp. 228–237. ICS 219, Excerpta Medica, Amsterdam.

Reimendal, R., and Sjövall, J. (1972): Analysis of steroids by off-line computerized gas chromatography-mass spectrometry. *Analytical Chemistry,* 44:21–29.

Samuelsson. B., Hamberg, M., and Sweeley, C. C. (1970): Quantitative gas chromatography of prostaglandin E_1 at the nanogram level. *Analytical Biochemistry,* 38:301–304.

Schomburg, G., and Henneberg, D. (1968): Zum Retentionsverhalten isotopenhaltiger Verbindungen unter Verwendung einer Isotopen-scan-Methode in einer Kombination Kapillargas-Chromatographie-Massenspektrometrie. *Chromatographia* 1:23–31.

Selke, E., Scholfield, C. R., Evans, C. D., and Dutton, H. J. (1961): Mass spectrometry and lipid research, *Journal of the American Oil Chemists' Society*, 38:614–615.

Sickmann, L., Hoppen, H.-O., and Breuer, H. (1970): Zur gaschromatographisch-massenspektrometrischen Bestimmung von Steroidhormonen in Körperflüssigkeiten unter Verwendung eines Multiple Ion Detectors (Fragmentographie). *Zeitschrift für Analytische Chemie*, 252:294–298.

Sjöquist, B., and Änggård, E. (1972): Gas chromatographic determination of homovanillic acid in human cerebrospinal fluid by electron capture detection and by mass fragmentography with a deuterated internal standard. *Analytical Chemistry*, 44:2297–2301

Story, M. S. (1972): Description of a GC/MS quadrupole instrument utilizing a chemical ionization source and no enriching device. In: *Twentieth Annual Conference on Mass Spectrometry and Allied Topics*, pp. 215–219. American Society for Mass Spectrometry, Dallas.

Strong, J. M., and Atkinson, Jr., A. J. (1972): Simultaneous measurement of plasma concentrations of lidocaine and its desethylated metabolite by mass fragmentography. *Analytical Chemistry*, 44:2287–2290.

Sweeley, C. C., Elliott, W. H., Fries, I., and Ryhage, R. (1966): Mass spectrometric determination of unresolved components in gas chromatographic effluents. *Analytical Chemistry*, 38:1549–1553

Sweetman, B. J., Frolich, J. C., and Watson, J. T. (1972): Conversion of PGE or PGE₂ to PGA or PGB as more suitable derivatives for vapor phase analysis into the subnanogram range. In: *Proceedings of the Fifth International Congress on Pharmacology*, p. 227. S. Karger, Basel.

Telling, G. M., Bryce, T. A., and Althorpe, J. (1971): Use of vacuum distillation and gas chromatography-mass spectrometry for determination of low levels of volatile nitrosamines in meat products. *Journal of Agricultural Food Chemistry*, 19:937–940.

Vance, D. E., and Sweeley, C. C. (1971): An *in vivo* study of the metabolism of human plasma glycosphingolipids using 6,6-dideuterioglucose as a tracer. In: *Proceedings of a Seminar on the Use of Stable Isotopes in Clinical Pharmacology*, edited by P. D. Klein and L. J. Roth. Argonne National Laboratory, National Technical Information Service, U.S. Department of Commerce, 5285 Port Royal Road, Springfield, Virginia 22151, Document number CONF-711115.

Vore, M., Gerber, N., and Bush, M. T. (1971): The metabolic fate of 1-n-butyl-5,5-diethylbarbiture acid in the rat. *Pharmacologist*, 13:220.

Walker, R. W., Ahn, H. S., Mandel, L. R., and VandenHeuvel, W. J. A. (1972): Identification of N,N-dimethyltryptamine as the product of an *in vitro* enzymic methylation. *Analytical Biochemistry*, 47:228–234.

Walle, T., and Gaffney, T. E. (1972): Propranolol metabolism in man and dog: Mass spectrometric identification of six new metabolites. *Journal of Pharmacology and Experimental Therapeutics*, 182:83–92.

Walle, T., Ishizaki, T., and Gaffney, T. E. (1973): Isopropylamine, a biologically active deamination product of propranolol in dogs: Identification of deuterated and unlabeled isopropylamine by gas chromatography-mass spectrometry. *Journal of Pharmacology and Experimental Therapeutics, in press*

Waller, G. R. (1972): *Biochemical Application of Mass Spectrometry*. Wiley-Interscience, New York.

Watson, J. T. (1971): The use of multiple ion detection in drug metabolism methodology. In: *Proceedings of a Seminar on the Use of Stable Isotopes in Clinical Pharmacology*, edited by P. D. Klein and L. J. Roth. Argonne National Laboratory, National Technical Information Service, U.S. Department of Commerce, 5285 Port Royal Road, Springfield, Virginia 22151, Document number CONF-711115.

Watson, J. T. (1972): Mass spectrometry instrumentation. In: *Biochemical Application of Mass Spectrometry*, edited by G. R. Waller. Wiley-Interscience, New York.

Watson, J. T., Pelster, D., Sweetman, B. J., and Frolich, F. C. (1972): Prostaglandin analysis with a GC-MS-computer system. In: *Twentieth Annual Conference on Mass Spectrometry and Allied Topics*, pp. 85–87. American Society for Mass Spectrometry, Dallas.

Watson, J. T., Sweetman, B. J., and Frolich, F. C. (1972): Quantification of prostaglandins at physiologically significant levels with GC-MS-computer system. In: *Proceedings of the Fifth International Congress on Pharmacology*, p. 248. S. Karger, Basel.

Wendt, G., and McCloskey, J. A. (1970): Mass spectrometry of perdeuterated molecules of biological origin. Fatty acid esters from "Scenedesmus obliquus." *Biochemistry*, 9:4854–4866.

Whitnack, E., Knapp, D. R., Holmes, J. C., Fowler, N. O., and Gaffney, T. E. (1972): Demethylation of nortriptyline by the dog lung. *Journal of Pharmacology and Experimental Therapeutics*, 181:288–291.

Wiesendanger, H. U. D., and Tao, F. T. (1970): Continuous monitoring of several compounds with a programmed mass spectrometer system. In: *Recent Developments in Mass Spectroscopy*, edited by K. Ogata and T. Haikawa, pp. 290-295. University of Tokyo Press, Japan.

Author Index

Subject Index

174